HUBBARD MODEL
AND
ANYON SUPERCONDUCTIVITY

Lecture Notes in Physics — Vol. 38

HUBBARD MODEL AND ANYON SUPERCONDUCTIVITY

A.P. Balachandran
Department of Physics, Syracuse University

E. Ercolessi
*Scuola di Perfezionamento in Fisica
Universita di Bologna and INFN*

G. Morandi
*Dipartimento di Fisica, Universita di Ferrara
INFM-CISM and INFN*

A.M. Srivastava
*Theoretical Physics Institute
University of Minnesota*

World Scientific
Singapore • New Jersey • London • Hong Kong

Published by

World Scientific Publishing Co. Pte. Ltd.
P O Box 128, Farrer Road, Singapore 9128
USA office: 687 Hartwell Street, Teaneck, NJ 07666
UK office: 73 Lynton Mead, Totteridge, London N20 8DH

Library of Congress Cataloging-in-Publication Data

Hubbard model and anyon superconductivity
    A. P. Balachandran . . . [et al.].
        p.   cm. -- (Lecture notes in physics; vol. 38)
    Includes bibliographical references.
    ISBN 9810203489. -- ISBN 9810203497 (pbk.)
    1. High temperature superconductivity -- Mathematical models.
2. Hubbard model.   I. Balachandran, A. P., 1938-   . II. Series:
Lecture notes in physics; 38.
QC611.98.H54H83   1990
537.6'23--dc20   90-45951
                                                                              CIP

Copyright © 1990 by World Scientific Publishing Co. Pte. Ltd.

All rights reserved. This book, or parts thereof, may not be reproduced in any form or by any means, electronic or mechanical, including photocopying, recording or any information storage and retrieval system now known or to be invented, without written permission from the Publisher.

Printed in Singapore by JBW Printers and Binders Pte. Ltd.

PREFACE

Several different models have recently been proposed to explain high temperature superconductivity.

This is a review of two such proposals, namely the Hubbard and anyon models. There is a belief that they are actually interrelated and not mutually exclusive, although a convincing demonstration of any such connection is yet to be found.

The choice of these topics as subjects of our review is dictated by our common interests. It is also dictated by our conviction of their physical importance regardless of their role in the study of the novel superconductors.

CONTENTS

Preface — v

Introduction — 1

1. The Hubbard Hamiltonian — 7

2. The Hubbard Hamiltonian in the Strong Coupling Limit — 13

3. Some Ideas for RVB — 24

4. RVB Ground States for the Strongly-Coupled Hubbard Hamiltonian at and below Half-Filling. Introduction to Superconductivity — 28

5. Other Mean Field Theories for the Hubbard-Heisenberg Hamiltonian on a 2D Square Lattice: Non-Uniform and Flux Phases — 39

6. Continuum Limit and the Chern-Simons Term — 63

7. The Abelian Chern-Simons Term — 73

8. The Non-Abelian Chern-Simons Term — 87

9. Anyon Superconductivity — 100

Acknowledgments — 110

References — 111

Appendix A. BCS and Projected BCS States — 116

Appendix B. A Cursory Look at the Quantum Hall Effect — 121

Appendix C. A Continuum Model For a System of Free Spins — 136

Introduction

The recent discovery of high-T_c superconducting materials [42, 78, 103] has revived interest in the physics of strongly correlated electronic systems [14, 16, 97, 106–108] and in the interplay between the metal-insulator transition (known as the "Mott transition" [138–140]) and the appearance of magnetic (typically antiferromagnetic) behaviors in such strongly correlated systems [52, 99, 100]. Qualitatively, the Mott transition is a cooperative, many-body phenomenon resulting from the competition between the tendency of electrons in a lattice to delocalize and to spread into energy bands and the mutual Coulomb repulsion, which tends to keep them apart and favors localization. For strong enough Coulomb repulsion (as compared with the energy gained by delocalizing the electrons into bands) complete localization may result, and systems which would be expected to behave as metals on the basis of a single-particle picture (for example, systems with an average number of one free electron per lattice site, like the vanadium oxides first studied by Mott [138–140]) exhibit instead insulating behavior.

Typically, the "new" superconductors are ceramic layered compounds with a highly complex unit cell (Fig. 1) of the perovskite type [41, 78, 103] which become superconducting upon doping, an almost paradigmatic example being $La_{2-x}(X)_xCuO_4$, with X = Ba or Sr. Pure La_2CuO_4 is considered to be a good example of a Mott insulator, and exhibits strong antiferromagnetic correlations (mainly in the Cu-O planes) as well. Stoichiometric La_2CuO_4 has (an average of) one electron per site. The further constraint

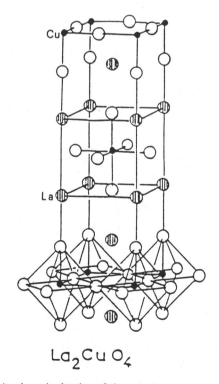

Fig. 1. A schematic drawing of the crystal structure of La_2CuO_4.

that, if the electron-electron correlations are strong enough, no two electrons should be allowed to simultaneously occupy the same site (at least at low enough temperatures and if one is only interested in the low-energy sector) leads to the stronger constraint that the ground state of the system should belong to the subspace of states which have exactly one electron per site. As will be seen later, but is also intuitively clear, states in such a subspace (which contains as many as 2^N states, with N the number of lattice sites) all correspond to an insulating behavior of the system. The above constraint has become widely known in the literature as the "half-filling" constraint (in the absence of correlations, one electron per site would correspond to a half-filled electronic band) or as the "Gutzwiller constraint", due to the fact that M. C. Gutzwiller was the first to employ it systematically in his investigations [88, 172] of strongly correlated electron systems.

The model Hamiltonian which is commonly agreed to best describe strongly correlated electrons bears the name of J. Hubbard, who first proposed it [106–108] (but see also the interesting introductory remarks in Ref. [19]). Some basic facts about the Hubbard Hamiltonian will be reviewed in Sec. 1 below.

The limit of strong correlations in the Hubbard Hamiltonian (in which the electron correlation energy (the (screened) Coulomb repulsion, which manifests itself when two electrons (of opposite spin) tend to occupy the same lattice site) is much greater than the bandwidth (of uncorrelated electrons)) has been investigated by various authors [3, 14, 16, 49, 50, 52, 65, 82, 97, 99, 100, 106–108, 149, 150, 166]. In particular, use has been made either of direct (degenerate) perturbation theory [65] or of canonical transformation methods [3, 49, 50, 82, 149, 150, 166]. All this material will be briefly reviewed in Sec. 2. The main result is that, to second-order in the kinetic-energy term (see Sec. 1 below for details) and at exact half-filling, the Hubbard Hamiltonian can be mapped exactly into an antiferromagnetic Heisenberg Hamiltonian, thus exhibiting clearly the dominant role of the magnetic correlations in the limit of strong Coulomb repulsion. In the cases of interest here, both Hamiltonians are supposed to describe essentially two-dimensional systems, the relevant dynamics taking place, in La-Cu oxides, essentially in the 2-D Cu-O layers. Interlayer couplings are much weaker than the intralayer ones, and do not seem to play a relevant dynamical role, although they can be important in stabilizing the superconducting transition [104].

The nature of the ground state of low-dimensionality (quantum) antiferromagnets is highly nontrivial, is still under debate and can be drastically different from that of the classical (ordered) Néel state [141, 142], even taking into account quantum corrections from spin-wave theory [134]. We recall that, in simple spin models with nearest-neighbor exchange, when the exchange favors antiparallel alignment of the spins, the classical ground state proposed by Néel [141, 142] consists of two interpenetrating sublattices (that is such that each spin in a sublattice has only nearest neighbors belonging to the other sublattice) each ordered ferromagnetically, and with the spins of the two sublattices pointing in opposite directions. This is possible for simple lattices (linear chains, cubic and honeycomb lattices in two dimensions, cubic, body-centered and face-centered cubic lattices in three dimensions, etc.). More complicated lattice structures can also be considered, increasing the number of sublattices. For example, for a plane triangular array of spins, one can divide the lattice into three interpenetrating sublattices (see Sec. 3

below), and the classical Néel state will then correspond to the spins forming mutual angles of 120°.

The one-dimensional antiferromagnetic Heisenberg chain was solved exactly in the early '30s by H. Bethe [44] using the famous "Bethe Ansatz" [44, 134, 171]. More recently, Lieb and Wu [127] have produced an exact solution for the ground state of the one-dimensional Hubbard model, also based on the Bethe Ansatz. We refer the interested reader to Refs. [134] and [171] for a more detailed discussion of the Bethe Ansatz, which will not be our main concern here mainly because its applicability is essentially restricted to actually [127, 171] or effectively [22] one-dimensional systems. In both of the cases mentioned above the ground state does not exhibit long-range order (which comes of course as no surprise, after all, since it is well known that one-dimensional systems cannot undergo phase transitions [122]), while the spins appear to be antiferromagnetically correlated in singlet pairs at short distances.

As early as in 1973, Anderson [16] suggested that a similar picture could apply to the description of two-dimensional antiferromagnets as well, namely that, compared with the classical Néel state, a better ground state could be obtained by allowing nearby spins to couple in singlet pairs. A little thought suffices to realize that such singlet bonds can exist, in any lattice, in a large variety of spatial configurations. Forming coherent superpositions of states corresponding to different bond configurations, that is allowing the bonds to "resonate" between different configurations, can lower the energy still further.

A detailed analysis on a specific model [68] proved this to be the case, and a ground-state energy substantially lower than that of the classical Néel state was found by numerically constructing better and better trial wave functions (i.e. by mixing more and more configurations) of the kind just described. Actually, when applied to the one-dimensional Heisenberg chain, a trial wave function of the type outlined above seems to yield a ground-state energy agreeing to within about 10% with the exact, Bethe-Ansatz, result, and this supports the validity of the picture. The proposed ground state seems therefore to be a sort of (quantum) liquid of singlet pairs resonating among various singlet configurations, whence the name of "Resonating Valence Bond" state ("RVB" for short) attributed to it.

In view of the correspondence discussed above between the Hubbard and the Heisenberg Hamiltonians, one is thus led to believe that the strong-coupling limit of the 2D Hubbard model at half-filling should be describable by an RVB-type ground state.

Anderson and coworkers [18, 21, 38, 41] tried to substantiate this idea quantitatively by developing a mean-field theory of the Hubbard Hamiltonian in the strong-coupling limit. The ground state proposed by the above authors turns out to be of the form of a doubly-projected BCS [156] state, the double projection coming from the constraint of a fixed number of particles and from the Gutzwiller constraint. In this description, the "Cooper pairs" are singlet bonds of nearest-neighbor electrons (or, more correctly [18], admixtures of singlet bonds of variable length). At exact half-filling, such singlet bonds can only resonate among different configurations, but are not free to move, and we have an insulating state.

Kivelson *et al.* [117] analyzed the structure of an RVB state on a 2D square lattice, and predicted the existence of peculiar topological elementary excitations, having the nature of neutral, spin-1/2 excitations roughly described as "dangling bonds", i.e. resulting from the breaking of a singlet pair. Such excitations are now known under the name of "spinons". The authors in Refs. [18] and [21, 38, 41] have identified the spinons with a class of quasiparticle excitations over the RVB-BCS-type ground state, namely excitations across a "pseudo-Fermi surface" which characterizes [18] the RVB ground state.

Doping, say, La_2CuO_4 by substituting trivalent Lanthanums with Sr or Ba drives the system below half-filling introducing holes, and the system becomes metallic. In the picture put forward in Ref. [117], the holes can bind to the spinons, thereby forming a gas of charged spinless bosons called "holons", and superconductivity is attributed to Bose condensation of the charged gas of holons. A related description of high-T_c superconductivity as Bose condensation of holes has been given in Ref. [104]. Such a picture meets however the difficulty that it assumes a singly-charged order parameter, while it appears experimentally well-founded [78, 103] that the order parameter of high-T_c superconductors (as well as that of the "ordinary", BCS ones) carries a charge which is twice the electron charge. Instead, in the RVB picture the system is, as already mentioned, metallic below half-filling. Thus, for example, [18] states the following: *"From a "mean-field" point of view, as soon as the system is metallized, it becomes a superconductor, since the pairing already exists in the RVB state . . ."*. Put in a nutshell, in the presence of holes the (charged) bound singlet pairs find space to move around, and this results in superconducting. This picture overcomes the difficulty mentioned above, and a rough estimate yields a superconducting transition temperature of the order of the binding energy of a singlet pair, which may well turn out to be of the right order of magnitude for typical high-T_c superconductors. However, the predicted dependence of the transition temperature on the concentration of dopant seems to fit rather badly [104] the existing experimental data.

As with every mean-field theory, in which fluctuations are averaged over, the approach of Anderson and coworkers can predict a genuine phase transition also in a system of low dimensionality. One should not forget, however, that a celebrated theorem in Quantum Statistical Mechanics, which goes under the names of Mermin and Wagner [136], states that, due to quantum fluctuations, no ODLRO (Off-Diagonal Long-Range Order) is possible in less than three space dimensions. That mean-field theories violate the Mermin-Wagner theorem is a well-known fact. Note that the authors of Ref. [104] take the weak interlayer coupling into account, and their transition temperature correctly vanishes when the interlayer coupling does and the system becomes truly two-dimensional. As already mentioned, the main role of the interlayer coupling seems to be that of stabilizing the transition.

When written in terms of the original Fermion operators, the Heisenberg Hamiltonian, into which the Hubbard Hamiltonian is mapped in the strong-coupling limit and at half-filling, turns out to admit gauge invariance both under local U(1) [39, 40, 47] and SU(2) [8, 55, 193] transformations, with the latter containing the former as a subgroup. This gauge invariance has led some authors [55] to investigate the connections between the description of a half-filled 2D Mott insulator and lattice QCD [118, 119], and also the mutual connections of various apparently different mean-field theories, among which is the "flux phase" of the RVB-BCS ground state [6, 20].

It has been known for quite some time [4, 5, 91–93] that in one space dimension the antiferrogmagnetic (AFM) Heisenberg chain can be mapped, in the continuum limit, into an effective field theory which turns out to be that of the O(3) nonlinear σ-model [151], and that a Wess-Zumino [180, 197] topological term appears in the effective Lagrangian in the process of taking the continuum limit. Such topological terms, ineffective at the classical level, are far from being irrelevant quantum mechanically whenever they are present in the Lagrangian. In particular, in three space dimensions [33] they can change the statistics of the excitations over the ground state, turning, e.g., nominal bosons into fermions (and vice versa). The situation is more complicated, but also far more fascinating [129, 182, 183, 185], in two space dimensions. We will resume this point shortly below.

The search for topological (actually Chern-Simons [58]) terms in the effective (low-energy, and hence also continuum-limit) Lagrangian of a 2D Heisenberg model has been pursued by various authors [110, 193], the reason for such a search being that Chern-Simons terms, if present, would put on rather firm grounds the more or less speculated existence of neutral spin excitations, i.e. the spinons, and would allow to determine their statistics.

Dzyaloshinskii, Polyakov and Wiegmann [63, 147] elaborated at some length on the consequences of the presence of Chern-Simons terms in the effective Lagrangian. However, the very existence of topological terms in two (space) dimensional theories has been questioned by various author [72, 175], and the problem is still under debate [11].

The peculiar role of topology in two-dimensional systems brings us to the last point we want to discuss in this introductory section.

It is well known [121] that the "canonical" statistics (Fermi or Bose) are the only possibilities of an assembly of identical particles in space dimension three, if we choose to ignore the possibility of parastatistics. On the contrary, the very concept of statistics of identical particles becomes ambiguous in one-dimensional systems [129]. Space dimension two represents instead an intermediate and very rich situation [101, 111, 126]. Indeed, the admissible statistics interpolate continuously between Bose and Fermi, and are classified by an angle θ (whence the name "θ-statistics"), with $\theta = 0$ and $\theta = \pi$ corresponding to Bose and Fermi statistics respectively. Identical particles in two dimensions are thus called "anyons" [77, 182].

A number of authors [25, 114, 196] have noticed that there is a close similarity between the RVB ground state and that proposed by Laughlin for the description of the Fractional Quantum Hall Effect (FQHE) [123]. Laughlin himself [124] has made a strong case that elementary excitations over the RVB ground state should exhibit a phenomenon of fractionalization of spin, just as elementary excitations in the FQHE exhibit fractionalization of charge [123]. It was then argued that, as a consequence, the elementary excitations should behave as "half-fermions", i.e. they should obey a "θ-statistics" with $\theta = \pi/2$. As, contrary to what could be a naive belief, pairs of half-fermions actually behave statistically as bosons, Laughlin [125] argued that superconductivity could result from Bose condensation of bound pairs of half-fermions. Further calculations [69] seem to support this point of view.

The problem of anyon superconductivity has been (and is being) also studied in a systematic way by many authors [51, 54, 69, 81, 98, 102]. A related problem is that of the so-called "Chiral Spin Liquids" [179], i.e. two-dimensional spin systems which can exhibit several (although related) types of order parameters, all of which violate both parity (P) and time-reversal (T). We recall that P- and T-violation also occurs in the FQHE, but this comes as no surprise in this case, and has indeed to be expected in the presence of a magnetic field. That the same situation may occur in the "flux phases" of an RVB quantum spin liquid had been suggested by Anderson [19, 20], and more recent theoretical work seems to support this point of view as well.

1. The Hubbard Hamiltonian

On quite general grounds, the Hamiltonian of an assembly of, say, N electrons on a given lattice can be written, in first-quantized language, as:

$$\mathscr{H} = \sum_i h(\mathbf{r}_i) + \frac{1}{2}\sum_{i\neq j} v(\mathbf{r}_i - \mathbf{r}_j) \tag{1.1}$$

where sums run from 1 to N, and $\mathbf{r}_i$ labels the position of the i-th electron, h is the "one-particle" part of the Hamiltonian (i.e. it contains the kinetic energy plus all the interactions with external potentials like the lattice potential and such) while v represents the electron-electron two-body interaction.

A convenient choice for an orthonormal basis of single-particle states is provided by the Wannier states [173, 192]. We recall that the Bloch functions appropriate to the description of the band structure of a solid are of the general form:

$$\psi_{n\mathbf{k}}(\mathbf{r}) = \exp[i\mathbf{k}\cdot\mathbf{r}]u_{n\mathbf{k}}(\mathbf{r}) \tag{1.2}$$

where n is a band index, $\mathbf{k}$ runs over the first Brillouin zone, and $u_{n\mathbf{k}}(\mathbf{r})$ has the periodicity of the lattice:

$$u_{n\mathbf{k}}(\mathbf{r} + \mathbf{R}_i) \equiv u_{n\mathbf{k}}(\mathbf{r}), \qquad \mathbf{R}_i \text{ a lattice vector}. \tag{1.3}$$

For a lattice with M sites, the Wannier functions $\phi_{ni}(\mathbf{r})$ are defined via the unitary transformation:

$$\phi_{ni}(\mathbf{r}) = \frac{1}{\sqrt{M}}\sum_{\mathbf{k}} \exp[-i\mathbf{k}\cdot\mathbf{R}_i]\psi_{n\mathbf{k}}(\mathbf{r}) \tag{1.4}$$

where the index (or multi-index) i labels the lattice site connected to the origin 0 by the lattice vector $\mathbf{R}_i$.

Using Eqs. (1.2) and (1.3), one easily obtains:

$$\phi_{ni}(\mathbf{r}) = \phi_{no}(\mathbf{r} - \mathbf{R}_i) \tag{1.5}$$

that is, one can define one Wannier function for each band, the others being obtained from it by translation. Being unitarily related to the Bloch functions, the Wannier functions form an orthonormal basis of single-particle functions:

$$\sum_{ni}\phi_{ni}(\mathbf{r})\phi_{ni}^*(\mathbf{r}') = \delta(\mathbf{r}-\mathbf{r}'); \qquad \langle\phi_{ni}|\phi_{mj}\rangle = \delta_{nm}\delta_{ij}. \tag{1.6}$$

As an example, let us consider a simple cubic lattice with lattice spacing d and, for a given band (whose index will be dropped), let us assume that $u_{n\mathbf{k}}$ has no appreciable $\mathbf{k}$-dependence: $u_{n\mathbf{k}}(\mathbf{r}) \simeq u(\mathbf{r})$. Then it is easy to show [192] that:

$$\phi_o(\mathbf{r}) = u(\mathbf{r}) \cdot \frac{\sin\left(\frac{\pi x}{d}\right)}{\left(\frac{\pi x}{d}\right)} \frac{\sin\left(\frac{\pi y}{d}\right)}{\left(\frac{\pi y}{d}\right)} \frac{\sin\left(\frac{\pi z}{d}\right)}{\left(\frac{\pi z}{d}\right)} \qquad (1.7)$$

ϕ_o is then centered around $\mathbf{r} = 0$ (and ϕ_i will be centered around $\mathbf{r} = \mathbf{R}_i$), and spreads out with gradually decreasing oscillations. This turns out to be a general feature of the Wannier functions, which are, in a sense, "the best one can do" to localize electrons in a lattice while retaining the orthonormality property of the set of single-particle wavefunctions.

As we will be considering one single band here, we will need just one Wannier function $\phi_o(\mathbf{r})$, centered at the origin, the others being given by:

$$\phi_j(\mathbf{r}) = \phi_o(\mathbf{r} - \mathbf{R}_j) . \qquad (1.8)$$

Introducing creation and annihilation operators $c^\dagger_{j\sigma}$ and $c_{j\sigma}$ for electrons in state ϕ_j and spin σ ($\sigma = \uparrow$ or $\downarrow$), the Hamiltonian (1.1) can be rewritten in the second quantized formalism as [70]:

$$\mathcal{H} = -\sum_{ij\sigma} t_{ij} c^\dagger_{i\sigma} c_{j\sigma} + \frac{1}{2} \sum_{ijkl} \sum_{\sigma\sigma'} \langle ij|v|kl\rangle c^\dagger_{i\sigma} c^\dagger_{j\sigma'} c_{l\sigma'} c_{k\sigma} \qquad (1.9)$$

where:

$$t_{ij} \equiv t(\mathbf{R}_i - \mathbf{R}_j) =: -\int d\mathbf{r}\, \phi_i^*(\mathbf{r}) h(\mathbf{r}) \phi_j(\mathbf{r}) = t_{ji}^* \qquad (1.10)$$

$$\langle ij|v|kl\rangle = \int d\mathbf{r} d\mathbf{r}'\, \phi_i^*(\mathbf{r}) \phi_j^*(\mathbf{r}') v(\mathbf{r} - \mathbf{r}') \phi_k(\mathbf{r}') \phi_l(\mathbf{r}) \qquad (1.11)$$

and both h and v have been assumed to be spin-independent (the appropriate generalizations, to include, e.g., spin-dependence, are easily done). Energies will be normalized in such a way that: $t_{ii} \equiv t(0) = 0$. The following approximations will now be made, which are however believed to retain the essential physics of strongly correlated electrons:

i) it will be assumed that $t_{ij} \equiv t(\mathbf{R}_i - \mathbf{R}_j)$ decays fairly rapidly with the distance, so that only matrix elements between nearest-neighbor sites need to be retained. For isotropic systems, we will then approximate t_{ij} as:

$$t_{ij} = \begin{cases} t & \text{for } (i, j) \text{ nearest neighbors ("n.n.")} \\ 0 & \text{otherwise} \end{cases} \qquad (1.12a)$$

The systems of interest are however, as discussed in the Introduction, layered compounds with an interlayer hopping integral which is believed to be substantially smaller than the intralayer one. Consequently, in this case we can slightly modify Eq. (1.12a) into [104]:

$$t_{ij} \simeq \begin{cases} t & \text{for } (ij) \text{ n.n. in a layer} \\ rt & \text{for } (ij) \text{ n.n. on n.n. layers} \\ 0 & \text{otherwise} \end{cases} \quad (1.12b)$$

with $r \ll 1$ giving the ratio of the interlayer-to-intralayer hopping strengths.

ii) The electron-electron Coulomb interaction is assumed to be effectively screened when electrons are far apart. The dominant contribution to the second term in Eq. (3) will then come from: $i = j = k = l$, i.e. when two electrons are on the same site, and we approximate:

$$\langle ij|v|kl\rangle \simeq \begin{cases} U & \text{if } i = j = k = l \\ 0 & \text{otherwise} \end{cases}. \quad (1.13)$$

The Pauli principle requires then: $\sigma' = -\sigma \equiv \bar{\sigma}$, and we thus eventually obtain the simplest (one-band) version of the Hubbard Hamiltonian:

$$\mathcal{H} = -\sum_{\langle ij\rangle\sigma} t_{ij} c^\dagger_{i\sigma} c_{j\sigma} + U \sum_i \hat{n}_{i\uparrow} \hat{n}_{i\downarrow} \quad (1.14)$$

where: $\hat{n}_{i\sigma} \equiv c^\dagger_{i\sigma} c_{i\sigma}$, the sums run over lattice sites, the symbol $\langle ij\rangle$ indicates a sum over nearest-neighbor pairs, and t_{ij} is given by Eqs. (1.12a) or (1.12b) as the case may be. For convenience, we make here the convention that the sum be made over ordered nearest-neighbor pairs, that is the pairs (i, j) and (j, i) will have to be counted once each. With this convention the summand will be symmetric under the interchange $i \Leftrightarrow j$. Sums over unordered pairs will be indicated with the subscript (i, j) instead of $\langle ij\rangle$ whenever necessary.

The limits $t \to 0$ and $U \to 0$ yield soluble Hamiltonians, namely:

i) $U \to 0$. The Hamiltonian is diagonalized by reverting from Wannier states to Bloch waves, namely by performing the Fourier transforms:

$$c_{i\sigma} = \frac{1}{\sqrt{M}} \sum_{\mathbf{k}} \exp[i\mathbf{k} \cdot \mathbf{R}_i] c_{\mathbf{k}\sigma} \quad (1.15)$$

where M is the number of lattice sites, and the sum over $\mathbf{k}$ runs over the first Brillouin zone [192]. This yields the diagonal Hamiltonian:

$$\mathcal{H}_{U=0} = \sum_{\mathbf{k}\sigma} \epsilon(\mathbf{k}) c^\dagger_{\mathbf{k}\sigma} c_{\mathbf{k}\sigma} \quad (1.16)$$

where the band energies are given by

$$\epsilon(\mathbf{k}) = -\sum_{\langle i,0\rangle} t_{i0} \exp[i\mathbf{k} \cdot \mathbf{R}_i] . \quad (1.17)$$

For example, for a two-dimensional square lattice with lattice constant a, and with t_{ij} given by Eq. (1.12a), we obtain:

$$\epsilon(\mathbf{k}) \equiv \epsilon(k_x, k_y) = -2t[\cos(k_x a) + \cos(k_y a)] .$$

$$|k_{x,y}| \leq \frac{\pi}{a} \tag{1.18}$$

The Brillouin zone and energy contours are displayed in Fig. 2 (taken from Hirsch in Ref. [100]) together with the position of the Fermi level for various band fillings. The total bandwidth is: $W = 8|t|$ in this case.

ii) $t = 0$. The Hamiltonian becomes:

$$\mathcal{H}_{t=0} = U \sum_i \hat{n}_{i\uparrow} \hat{n}_{i\downarrow} . \tag{1.19}$$

The electrons have no dynamics any more and, for $U > 0$ (the case of interest to us) they will tend, in the ground state, to avoid double occupancies as much as possible. In particular, at half filling ($N = M$) the ground state will contain exactly one electron per site, with degeneracy 2^N (in the absence of external magnetic fields). It is quite plausible that, even switching on the hopping term (i.e. the first term in Eq. (1.14)), hopping electrons from one site to a nearby one will cost high in energy (of order U), and therefore the hopping term will remain ineffective as long as $|t| \ll U$. The low-energy sector of the model will remain insulating even when $t \neq 0$, and we have an elementary (and certainly oversimplified) description of a Mott insulator.

For intermediate couplings, the model is not exactly soluble, except in one space dimension [127]. All the statistical information can be obtained from the partition function:

$$\mathcal{Z} = \text{Tr}[\![\exp(-\beta \mathcal{H})]\!] . \tag{1.20}$$

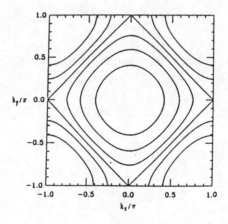

Fig. 2. The Brillouin zone of a 2D tight-binding model. Contours indicate the shape of the Fermi surface for various fillings.

Writing the Hamiltonian as: $\mathcal{H} = \mathcal{H}_o + V$, with $\mathcal{H}_o$ the hopping term and $V = U \sum_i \hat{n}_{i\uparrow} \hat{n}_{i\downarrow}$, Eq. (1.20) can be rewritten as:

$$\mathcal{Z} = \mathcal{Z}_o \left\langle T_\tau \exp\left[-\int_0^\beta d\tau V(\tau)\right]\right\rangle_o \quad (1.21)$$

where: $\mathcal{Z}_o := \text{Tr} \exp[-\beta\mathcal{H}_o]$, $\langle\langle(\ldots)\rangle\rangle = \mathcal{Z}_o^{-1} \text{Tr}\{\exp[-\beta\mathcal{H}_o](\ldots)\}$, T_τ is a "time" ordering operator in the imaginary-time (that is for: $t = -i\tau$) domain and: $V(\tau) = \exp(-\tau\mathcal{H}_o)V\exp(+\tau\mathcal{H}_o)$. Equation (1.21) is simply the interaction-representation expression for the partition function, written in the imaginary-time domain, and is a standard starting point for thermodynamic perturbation theory [70].

Using the Stratonovich-Hubbard identity [105, 170], the two-body interaction V can be linearized to yield the functional-integral expression [137]:

$$\mathcal{Z} = \mathcal{Z}_o \int \left(\prod_i \mathcal{D}x_i(\tau)\mathcal{D}y_i(\tau)\right) \exp\left[-\pi \sum_j \int_0^1 d\tau(x_j(\tau)^2 + y_j(\tau)^2)\right]$$

$$\times \left\langle T_\tau \exp\left[-\sum_{j\sigma}\int_0^1 d\tau v_{j\sigma}(\tau)\hat{n}_{j\sigma}(\tau)\right]\right\rangle_o \quad (1.22)$$

for the partition function. Here:

$$v_{j\sigma}(\tau) = c[\sigma x_j(\tau) + iy_j(\tau)] ; \qquad \sigma = \pm 1 ; \qquad c := \sqrt{\pi\beta U} . \quad (1.23)$$

The product of the unperturbed partition function $\mathcal{Z}_o$ and of the expectation value in Eq. (1.22) is the partition function of a gas of free electrons interacting with "time" and spin-dependent potentials $v_{j\sigma}(\tau)$. The latter (or, better, the fields $x_j(\tau)$ and $y_j(\tau)$) are independent random variables with a Gaussian distribution determined by the first exponential in Eq. (1.22). Each Gaussian measure is normalized:

$$\int \mathcal{D}z(\tau) \exp\left[-\pi \int_0^1 d\tau z^2(\tau)\right] = 1 ; \qquad z = x_j \text{ or } z = y_j \quad (1.24)$$

has zero mean and variance [137]:

$$f(\tau, \tau') =: \int \mathcal{D}z(\tau'') \exp\left[-\pi \int_0^1 d\tau'' z^2(\tau'')\right] z(\tau)z(\tau') = \delta(\tau - \tau') . \quad (1.25)$$

Moreover, one can prove [137] the following Gaussian analogue of Wick's theorem, namely:

$$\int \mathcal{D}z(\tau) \exp\left[-\pi \int_0^1 d\tau z^2(\tau)\right] z(\tau_1) \ldots z(\tau_{2k})$$

$$= \frac{1}{(2\pi)^k} \sum_{\text{perm}} f(\tau_1, \tau_2) \ldots f(\tau_{2k-1}, \tau_{2k}) \quad (1.26)$$

where the sum ranges over all the possible distinct permutations of $\tau_1 \ldots \tau_{2k}$, while the integrals of odd monomials against the Gaussian measure all vanish. By expanding the second exponential in Eq. (1.22), the resulting Gaussian integrations can all be performed, and one can prove [137] that the standard perturbation expansion (in U) can be recovered in this way. More details on the functional-integral approach can be found in Ref. [137] (and in the references quoted therein).

The expectation value in Eq. (1.22) can be evaluated either by Grassmann integration over the fermionic variables, or by direct evaluation of the trace [137]. The result is a functional integral involving a fermion determinant which is a highly nonlinear and nonlocal expression in the random fields $v_{j\sigma}$. We will not pursue this calculation here, but will come back to a closely related one in a later section. Cyrot [52] resorted to a saddle-point evaluation of the path-integral. This is equivalent [105, 170] to performing the following Hartree-Fock factorization of the original Hamiltonian:

$$\mathcal{H} \simeq - \sum_{\langle i,j \rangle \sigma} t_{ij} c^\dagger_{i\sigma} c_{j\sigma} + \frac{1}{2} U \sum_{i\sigma} \langle \hat{n}_{i\sigma} \rangle \hat{n}_{i\bar{\sigma}} - U \sum_i \langle \hat{n}_{i\uparrow} \rangle \langle \hat{n}_{i\downarrow} \rangle . \quad (1.27)$$

Here, at the end of the calculations, the expectation values $\langle n_{i\sigma} \rangle$ have to be evaluated self-consistently. We refer to the original paper of Cyrot [52] for a discussion of the three-dimensional case, and report in Fig. 3 the phase diagram at $T = 0$ obtained, in the same approximation, by Hirsch [100] in the two-dimensional case.

The high nonlinearity of the self-consistency equations leads to the rich variety of magnetic (or non-magnetic) behaviors of the models as the physical parameter $U/|t|$ (or U/W, with W the bandwidth) is varied. From Fig. 3 we see in particular that, from small values of $U/|t|$ onwards, the system is an antiferromagnetic insulator at half-filling. This confirms (only at the mean-field level, though) that the Hubbard Hamiltonian predicts the Mott transition in the strong-coupling limit and at half-filling. Quite similar results were obtained by Cyrot in the three-dimensional case. The strong-coupling limit of the Hubbard Hamiltonian will be the topic of the next section.

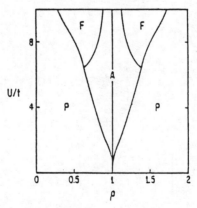

Fig. 3. Phase diagram for a 2D Hubbard model on a square lattice in mean-field theory. A = Antiferromagnetic; F = Ferromagnetic; P = Paramagnetic.

2. The Hubbard Hamiltonian in the Strong Coupling Limit

By "strong-coupling" limit of the Hamiltonian of Eq. (1.14) we mean the leading-term expansion in $|t|/U$, which is the natural "small parameter" of the theory when $|t| \ll U$. We write again:

$$\mathcal{H} = \mathcal{H}_o + V \tag{2.1}$$

where $\mathcal{H}_o$ is the hopping term and $V = U \sum_i \hat{n}_{i\uparrow} \hat{n}_{i\downarrow}$. The eigenvalues of V (the $t \to 0$ limit of the full Hubbard Hamiltonian) are given by:

$$E_l^{(o)} = l \cdot U \tag{2.2}$$

where l is the number of doubly-occupied sites, $0 \leq l \leq M$ (M being the number of lattice sites). Hence, $\eta := l/M$ represents the probability of any site being doubly occupied. If N ($\leq 2M$) is the total number of electrons, the ground state, for $N \leq M$, corresponds to $l = 0$ ($\eta = 0$), and has degeneracy:

$$D_0 = 2^N \cdot \begin{bmatrix} M \\ N \end{bmatrix} \tag{2.3a}$$

where the first factor comes from the fact that, for a given distribution of occupied sites, there are two possible spin orientations for each site, and the second factor (a binomial coefficient) counts the way of distributing N (indistinguishable) objects among M sites. If instead $N \geq M$, the ground state will correspond to: $l = N - M$ ($\eta = (N/M) - 1$), and the degeneracy of the ground state will be:

$$D_0 = 2^{2M-N} \cdot \begin{bmatrix} M \\ 2M - N \end{bmatrix} \tag{2.3b}$$

($2M - N$ being the number of singly-occupied sites in the ground state).

Projections onto the subspaces $\mathcal{H}_l$ of the full Hilbert space corresponding to $E_l^{(o)}$ for the various values of l can be obtained in a compact form from the expansion of:

$$\Pi(x) := \prod_{i=1}^{M} [1 - (1 - x)\hat{v}_i] \; ; \qquad \hat{v}_i := \hat{n}_{i\uparrow} \hat{n}_{i\downarrow} \; ; \qquad 0 \leq x \leq 1 \; . \tag{2.4}$$

Indeed, defining the expansion coefficients P_l via:

$$\Pi(x) := \sum_{l=0}^{M} x^l P_l \tag{2.5}$$

one easily obtains:

$$P_l = \sum_{(i_1 \ldots i_l)} \left[\hat{v}_{i_1} \ldots \hat{v}_{i_l} \prod_j{}' (1 - \hat{v}_j) \right] \tag{2.6}$$

where the sum stands for a sum over all unordered l-tuples of (pairwise distinct) indices, and $\prod'_j$ stands, for any l-tuple, for a product over all the j's not in the given l-tuple. One can easily convince oneself that P_l is the required projection onto the subspace with l doubly occupied sites. In particular:

$$P_o = \prod_{i=1}^{M} (1 - \hat{v}_i) \tag{2.7}$$

projects onto the subspace containing no doubly occupied sites at all, and is known, and usually quoted as, the "Gutzwiller projection". Note also that: $\Pi(1) = \sum_l P_l = 1$ as it should be.

The expansion in $|t|/U$ has been pursued in the literature in two essentially equivalent ways: either by direct perturbation expansions [65] (using Kato's [115] resolvent formalism and degenerate perturbation theory) or by the use [3, 49, 50, 82, 104, 149, 150, 166] of a suitable canonical transformation, a technique reminiscent of the Schrieffer-Wolff transformation which was devised as early as in 1965 by Schrieffer and Wolff [157] to treat the strong coupling limit of the single impurity Anderson model [15]. We will follow the latter line of approach here.

Defining (cf. Eq. (2.6)):

$$P_\eta := \sum_{l>0} P_l \tag{2.8}$$

P_η will project onto the subspace containing at least one doubly occupied site, and, of course:

$$P_0 + P_\eta = 1 \,. \tag{2.9}$$

On physical grounds, the decomposition of the Hilbert space into the orthogonal sum of the subspaces spanned by P_0 and P_η corresponds to the fact that, for U large (on the scale of the bandwidth, i.e., roughly speaking, of $|t|$), the latter is expected to be separated by a large energy gap from the former, which will then describe the low-energy sector of the model at and below half-filling ($N \leq M$).

Using the decomposition of the identity (2.9), we can write:

$$\mathcal{H}_0 \equiv (P_0 + P_\eta)\mathcal{H}_0(P_0 + P_\eta) = P_0\mathcal{H}_0P_o + P_\eta\mathcal{H}_0P_\eta + P_0\mathcal{H}_0P_\eta + P_\eta\mathcal{H}_0P_0 \tag{2.10}$$

while, of course:

$$V \equiv P_\eta V P_\eta \tag{2.11}$$

(i.e. $P_0 V P_0 = P_\eta V P_0 = 0$).

Note that, if the decomposition (2.8) is used for P_η, one has: $P_l \mathcal{H}_0 P_{l'} = 0$ unless: $\Delta l = l - l' = 0, \pm 1$, while: $P_l V P_{l'} = \delta_{ll'} P_l V P_l$. We will not need however this finer decomposition in what follows, except at the very end of the calculations.

Before analyzing the various terms in the r.h.s. of Eq. (2.10), let's observe that the use of the (trivial) identity:

$$c_{i\sigma} \equiv c_{i\sigma}[(1 - \hat{n}_{i\bar{\sigma}}) + \hat{n}_{i\bar{\sigma}}] \tag{2.12}$$

allows us to rewrite $\mathcal{H}_0$ as:

$$\mathcal{H}_0 \equiv T_h + T_d + T_{\text{mix}} \tag{2.13}$$

where:

$$T_h = - \sum_{\langle ij \rangle \sigma} t_{ij}(1 - \hat{n}_{i\bar{\sigma}})c^\dagger_{i\sigma} c_{j\sigma}(1 - \hat{n}_{j\bar{\sigma}}) \; ; \qquad \bar{\sigma} = -\sigma \tag{2.14}$$

$$T_d = - \sum_{\langle ij \rangle \sigma} t_{ij} \hat{n}_{i\bar{\sigma}} c^\dagger_{i\sigma} c_{j\sigma} \hat{n}_{j\bar{\sigma}} \tag{2.15}$$

$$T_{\text{mix}} = - \sum_{\langle ij \rangle \sigma} \{t_{ij} \hat{n}_{i\bar{\sigma}} c^\dagger_{i\sigma} c_{j\sigma}(1 - \hat{n}_{j\bar{\sigma}}) + \text{h.c.}\} \; . \tag{2.16}$$

One can check by inspection that:
i) Both T_h and T_d preserve the number of doubly occupied sites, i.e. they commute with V. Indeed, T_h transfers one electron from a singly occupied site to an empty one, while T_d acts on a doubly occupied site, transferring one electron to a previously singly occupied one, i.e. it destroys one electron pair on a site recreating it on a nearest-neighbor one. All this implies of course:

$$P_0 T_h P_\eta = P_0 T_d P_\eta = 0 \; . \tag{2.17}$$

Moreover, T_d (as well as V) vanishes in the subspace spanned by the Gutzwiller projection:

$$P_0 T_d P_0 = 0 \tag{2.18}$$

while, however:

$$P_\eta T_h P_\eta \neq 0 \tag{2.19}$$

(single-electron transfer between singly-occupied and empty sites can take place also in the presence of other doubly-occupied sites). However, T_h will vanish on the states, also belonging to the Gutzwiller projection, in which $M = N$, i.e. at exact half-filling.
ii) T_{mix} is the only term which can mix different eigenspaces of V.

A digression on: "Projected" operators and their algebra

The annihilation operators $c_{i\sigma}$ can be decomposed in an obvious way as:

$$c_{i\sigma} \equiv c_{i\sigma}[(1 - \hat{\nu}_i) + \hat{\nu}_i] \ . \tag{2.20}$$

We recall that $(1 - \hat{\nu}_i)$ is the projection operator onto the states in which site i is either empty or singly occupied, while $\hat{\nu}_i$ projects onto the double occupancy of the same site. Using then the identity:

$$c_{i\sigma}\hat{n}_{i\sigma} \equiv c_{i\sigma} \quad \forall i, \sigma \tag{2.21}$$

we can define the "projected" operators

$$d_{i\sigma} := c_{i\sigma}(1 - \nu_i) \equiv c_{i\sigma}(1 - \hat{n}_{i\bar{\sigma}}) \tag{2.22}$$

$$\tilde{d}_{i\sigma} := c_{i\sigma}\nu_i \equiv c_{i\sigma}\hat{n}_{i\bar{\sigma}} \tag{2.23}$$

together with their adjoints. Equations (2.14)–(2.16) can be rewritten as:

$$T_h = \sum_{\langle ij \rangle \sigma} t_{ij} d_{i\sigma}{}^\dagger d_{j\sigma} \tag{2.14'}$$

$$T_d = -\sum_{\langle ij \rangle \sigma} t_{ij} \tilde{d}_{i\sigma}{}^\dagger \tilde{d}_{j\sigma} \tag{2.15'}$$

$$T_{\text{mix}} = -\sum_{\langle ij \rangle \sigma} [t_{ij} \tilde{d}_{i\sigma}{}^\dagger d_{j\sigma} + \text{h.c.}] \ . \tag{2.16'}$$

However, the projected operators are not true fermion operators. Indeed, one finds, with some simple algebra:

$$\{d_{i\sigma}, d_{j\tau}{}^\dagger\} = (1 - \hat{n}_{i\bar{\sigma}})\delta_{ij}\delta_{\sigma\tau} + \delta_{ij}\delta_{\sigma\bar{\tau}} c_{i\bar{\sigma}}{}^\dagger c_{i\sigma} \tag{2.24}$$

where the symbol $\{.,.\}$ stands for an anticommutator, and: $\hat{n}_i = \hat{n}_{i\uparrow} + \hat{n}_{i\downarrow}$. Similarly, one finds:

$$\{\tilde{d}_{i\sigma}, \tilde{d}_{j\tau}{}^\dagger\} = \delta_{ij}\delta_{\sigma\tau}\hat{n}_{i\bar{\sigma}} - \delta_{ij}\delta_{\sigma\bar{\tau}} c_{i\bar{\sigma}}{}^\dagger c_{i\sigma} \ . \tag{2.25}$$

The only remaining anticommutator is between a d and a $\tilde{d}$. It can be deduced from:

$$\{c_{i\sigma}, c_{j\tau}{}^\dagger\} \equiv \{d_{i\sigma} + \tilde{d}_{i\sigma}, d_{j\tau}{}^\dagger + \tilde{d}_{j\tau}{}^\dagger\} = \delta_{ij}\delta_{\sigma\tau} \tag{2.26}$$

and turns out to be zero.

Returning now to the projected decomposition of $\mathcal{H}_0$ of Eq. (2.10), one can easily convince oneself that the various terms of the decomposition are given by:

$$P_0 \mathcal{H}_0 P_0 = P_0 T_h P_0 \tag{2.27}$$

$$P_\eta \mathcal{H}_0 P_\eta = P_\eta T_h P_\eta + T_d \tag{2.28}$$

$$P_0 \mathcal{H}_0 P_\eta + P_\eta \mathcal{H}_0 P_0 = T_{\text{mix}} \ . \tag{2.29}$$

Note that, in view of what has been been said before that after Eq. (2.16) (see comment i)) T_d and T_{mix} in Eqs. (2.28)–(2.29) need not be "sandwiched" between projection operators.

Defining now:

$$\mathcal{H} = \tilde{\mathcal{H}}_0 + \mathcal{H}_\eta \qquad (2.30)$$

where:

$$\tilde{\mathcal{H}}_0 := P_0 \mathcal{H}_0 P_0 + P_\eta \mathcal{H}_0 P_\eta + V \qquad (2.31)$$

$$\mathcal{H}_\eta := P_0 \mathcal{H}_0 P_\eta + P_\eta \tilde{\mathcal{H}}_0 P_0 \qquad (2.32)$$

the Hubbard Hamiltonian is decomposed into the sum of a "diagonal" part $\tilde{\mathcal{H}}_0$, and an "off-diagonal" part $\mathcal{H}_\eta$ coupling the subspaces spanned by P_0 and P_η.

We now look for a canonical transformation eliminating the effect of $\mathcal{H}_\eta$ to lowest order, i.e. such that the transformed Hamiltonian $\mathcal{H}_{\text{eff}}$ is completely of the "diagonal" form. This will be embodied in the requirement that $\mathcal{H}_{\text{eff}}$ satisfies:

$$P_0 \mathcal{H}_{\text{eff}} P_\eta = 0 \qquad (2.33)$$

to the required order.

Let us proceed formally, and define:

$$\mathcal{H}(\epsilon) := \tilde{\mathcal{H}}_0 + \epsilon \mathcal{H}_\eta \qquad (2.34)$$

in terms of a parameter ϵ, with $\epsilon = 1$ corresponding to the physical case. We then seek a canonical transformation in the form:

$$\mathcal{U}(\epsilon) = \exp[i\epsilon S] \; ; \qquad S = S^\dagger \qquad (2.35)$$

in such a way that the transformed Hamiltonian $\mathcal{H}_{\text{eff}}(\epsilon)$ obeys:

$$\mathcal{H}_{\text{eff}} := \exp[i\epsilon S] \mathcal{H}(\epsilon) \exp[-i\epsilon S] = \tilde{\mathcal{H}}_0 + \mathcal{O}(\epsilon^2) \qquad (2.36)$$

i.e. we seek to eliminate the effects of $\mathcal{H}_\eta$ to order ϵ. As, at the end of the calculations, we have to set $\epsilon = 1$, it is clear that ϵ is not an expansion parameter of any sort, but simply a bookkeeping device for keeping track consistently of all the terms in $\mathcal{H}$ we wish to get rid of. Expanding Eq. (2.36), we find:

$$\mathcal{H}_{\text{eff}}(\epsilon) = \tilde{\mathcal{H}}_0 + \epsilon[\mathcal{H}_\eta + i[S, \tilde{\mathcal{H}}_0]] + \epsilon^2(i[S, \mathcal{H}_\eta] + \tfrac{1}{2}[S, [\tilde{\mathcal{H}}_0, S]]) + \mathcal{O}(\epsilon^3) \,. \qquad (2.37)$$

The condition determining the generator S is then:

$$[\tilde{\mathcal{H}}_0, S] + i \mathcal{H}_\eta = 0 \qquad (2.38)$$

and, setting $\epsilon = 1$, we eventually find:

$$\mathcal{H}_{\text{eff}} = \tilde{\mathcal{H}}_0 + \frac{i}{2} [S, \mathcal{H}_\eta] \tag{2.39}$$

with S obeying Eq. (2.38).

Before we start discussing the solution(s) of Eq. (2.38), let us observe that, with $\mathcal{H}_{\text{eff}}$ given by Eq. (2.39), and by repeated use of the decomposition of the identity (2.9), one easily obtains:

$$P_0 \mathcal{H}_{\text{eff}} P_0 = P_0 \mathcal{H} P_0 + \frac{i}{2} [(P_0 S P_\eta)(P_\eta \mathcal{H} P_0) - (P_0 \mathcal{H} P_\eta)(P_\eta S P_0)] \tag{2.40}$$

and

$$P_\eta \mathcal{H}_{\text{eff}} P_\eta = P_\eta \mathcal{H} P_\eta + \frac{i}{2} [(P_\eta S P_0)(P_0 \mathcal{H} P_\eta) - (P_\eta \mathcal{H} P_0)(P_0 S P_\eta)] \tag{2.41}$$

while:

$$P_\eta \mathcal{H}_{\text{eff}} P_0 = \frac{i}{2} [(P_\eta S P_\eta)(P_\eta \mathcal{H} P_0) - (P_\eta \mathcal{H} P_0)(P_0 S P_0)] . \tag{2.42}$$

This is a remarkable result, showing that:

i) The diagonal parts of $\mathcal{H}_{\text{eff}}$ are determined solely by the off-diagonal parts of the generator S, and

ii) The diagonal parts of S determine the off-diagonal part of $\mathcal{H}_{\text{eff}}$, and have therefore to be determined in such a way as to fulfill Eq. (2.33).

Returning now to Eq. (2.38), and using again Eq. (2.9), we find, after some algebra, the equivalent set of equations:

$$P_0 [\tilde{\mathcal{H}}_0, S] P_0 = 0 \Rightarrow [P_0 \mathcal{H} P_0, P_0 S P_0] = 0 \tag{2.43}$$

$$P_\eta [\tilde{\mathcal{H}}_0, S] P_\eta = 0 \Rightarrow [P_\eta \mathcal{H} P_\eta, P_\eta S P_\eta] = 0 \tag{2.44}$$

and

$$P_0 [\tilde{\mathcal{H}}_0, S] P_\eta + i P_0 \mathcal{H}_\eta P_\eta = 0 \Rightarrow (P_0 S P_\eta)(P_\eta \mathcal{H} P_\eta) - (P_0 \mathcal{H} P_0)(P_0 S P_\eta) = i P_0 \mathcal{H} P_\eta \tag{2.45}$$

Choosing then:

$$P_0 S P_0 = \lambda P_0 ; \quad P_\eta S P_\eta = \lambda P_\eta \tag{2.46}$$

for arbitrary (real) λ allows us to satisfy simultaneously Eqs. (2.43)–(2.44) and (cf. Eq. (2.42)) Eq. (2.33) as well. As λ remains entirely arbitrary, we will choose $\lambda = 0$ in what follows. For more general solutions, see Refs. [50, 149, 150].

We are thus left with the last equation, (2.45), to determine the off-diagonal elements of S.

Let us denote by $\mathcal{E}_0$ the subspace spanned by the Gutzwiller projection P_0, and by $\mathcal{E}_\eta$ its orthogonal complement (i.e. the subspace spanned by P_η). Also, to simplify the notation, let's define:

$$X := P_0 S P_\eta \tag{2.47a}$$

$$Y := P_\eta \mathcal{H} P_\eta \tag{2.47b}$$

$$R := P_0 \mathcal{H} P_\eta \tag{2.47c}$$

$$Q := P_0 \mathcal{H} P_0 \tag{2.47d}$$

Then, Eq. (2.45) becomes:

$$X \cdot Y - Q \cdot X = iR \tag{2.48}$$

and is a linear operator equation for the unknown X. Now, the operator Y has obviously no inverse in the full Hilbert space of states $\mathcal{E}_0 \oplus \mathcal{E}_\eta$. However, in the limit we are considering in this section, it is plausible to assume that Y is actually positive definite (due to the large U-term) in $\mathcal{E}_\eta$, and can then be inverted in that subspace. Calling then, with some abuse of notation, Y^{-1} the inverse of Y in the above subspace, we can rewrite Eq. (2.48) as:

$$X = iR \cdot Y^{-1} + Q \cdot X \cdot Y^{-1} . \tag{2.48'}$$

Equation (2.48') has then the solution (defined under reserve of convergence):

$$X = i \sum_{n=0}^{\infty} Q^n \cdot R \cdot Y^{-n-1} . \tag{2.49}$$

Now, in the limit ($|t| \ll U$) we are considering here, each one of the sharp levels of: $V = U \sum_i \hat{n}_{i\uparrow} \hat{n}_{i\downarrow}$ is broadened into a band of total width W ($\propto |t|$) by the effect of the hopping term $\mathcal{H}_0$, and neighboring bands are widely separated by an energy of order $U \gg W$. The order of magnitude of $Q \cdot Y^{-1}$ is therefore $|t| \cdot U^{-1}$, and Eq. (2.49) is actually an expansion in such a parameter. Approximating then [49, 50] Y by: $U \cdot P_\eta$ (and Y^{-1} by $U^{-1} \cdot P_\eta$), we obtain, to lowest order:

$$X \equiv P_0 S P_\eta = iR \cdot Y^{-1} \equiv \frac{i}{U} P_0 \mathcal{H} P_\eta \tag{2.50}$$

with (cf. Eqs. (2.29) and (2.16)):

$$P_0 \mathcal{H} P_\eta = - \sum_{\langle ij \rangle \sigma} t_{ij}(1 - \hat{n}_{i\bar{\sigma}}) c^\dagger_{i\sigma} c_{j\sigma} \hat{n}_{j\bar{\sigma}} . \qquad (2.51)$$

Inserting the result (2.50) into Eqs. (2.40)–(2.41), we find:

$$P_0 \mathcal{H}_{\text{eff}} P_0 = P_0 \mathcal{H} P_0 - \frac{1}{U} P_0 \mathcal{H} P_\eta \mathcal{H} P_0 \qquad (2.52)$$

and

$$P_\eta \mathcal{H}_{\text{eff}} P_\eta = P_\eta \mathcal{H} P_\eta + \frac{1}{U} P_\eta \mathcal{H} P_0 \mathcal{H} P_\eta . \qquad (2.53)$$

We will concentrate here on the effective Hamiltonian in the low-energy sector, i.e. Eq. (2.52). Let us note parenthetically that, in view of the selection rule discussed right after Eq. (2.11), it is only P_1, the projector corresponding to $l = 1$, and not the full projector P_η which is effective in Eq. (2.52).

A digression on: on-site spin operators

For further use, we introduce now the vector operators:

$$\mathbf{S}_i := \frac{1}{2} \sum_{\sigma \sigma'} c^\dagger_{i\sigma} [\boldsymbol{\sigma}]_{\sigma \sigma'} c_{i\sigma'} \qquad (2.54)$$

where: $\boldsymbol{\sigma} \equiv (\sigma^1, \sigma^2, \sigma^3)$ is the set of Pauli matrices. Explicitly:

$$S_i^+ := S_i^1 + i S_i^2 = c^\dagger_{i\uparrow} c_{i\downarrow} ; \qquad S_i^- := (S_i^+)^\dagger \qquad (2.55)$$

and

$$S_i^3 := \frac{1}{2} (\hat{n}_{i\uparrow} - \hat{n}_{i\downarrow}) . \qquad (2.56)$$

It is easy to check that the operators (2.55)–(2.56) close on the Lie algebra of SU(2).

Considering the (four-state) Hilbert space relative to site i (empty, singly occupied (two spin states) and doubly occupied states), the operators (2.55)–(2.56) act irreducibly on the subspace of singly occupied states, and trivially (i.e. annihilating them) on empty and doubly occupied states, i.e. they act as spin-zero operators on both of the above states. In the (two-dimensional) subspace in which site i is singly occupied, the operators (2.55)–(2.56) act instead as spin-1/2 operators. On the other hand, in the subspace in which site i is either doubly occupied or empty, the same SU(2) algebra is realized in a non-trivial way by:

$$\tilde{S}_i^+ := c^\dagger_{i\uparrow} c^\dagger_{i\downarrow} \; ; \qquad \tilde{S}_i^- = (\tilde{S}_i^+)^\dagger \tag{2.55'}$$

$$\tilde{S}_i^3 := \frac{1}{2}(\hat{n}_i - 1) \tag{2.56'}$$

(with the doubly occupied state acting as "spin-up" and the empty state as "spin-down" states respectively). The two sets of operators are actually connected by the following canonical transformations:

$$c_{i\uparrow} \to \tilde{c}_{i\uparrow} := c_{i\uparrow} \tag{2.57a}$$

$$c_{i\downarrow} \to \tilde{c}_{i\downarrow} := c^\dagger_{i\downarrow} \tag{2.57b}$$

i.e. by a particle-hole transformation in one spin channel.

That the operators (2.55)–(2.56) correspond (only) to $S = 0, 1/2$ can also be seen by constructing the Casimir operator, i.e. $\mathbf{S}^2$. It turns out that:

$$\mathbf{S}^2 \equiv 3 \cdot (S^3)^2 \tag{2.58}$$

$\mathbf{S}^2$ does indeed commute with all the S^i's, as, e.g.:

$$[\![\mathbf{S}^2, S^+]\!] \equiv 3 \cdot S^3 \cdot \{S^3, S^+\} \tag{2.59}$$

and the anticommutator on the r.h.s. vanishes identically ($S^3 S^+ = S^+/2$, $S^+ S^3 = -S^+/2$). On every subspace on which the algebra of operators acts irreducibly, $\mathbf{S}^2$, and hence $(S^3)^2$, must be a multiple of the identity, and this will be possible iff $S^3 = 0$ or $S^3 = \pm 1/2$. □

Let us turn back now to Eq. (2.52). Due to the presence of the various projection operators, $P_\eta \mathcal{H} P_0 \equiv P_1 \mathcal{H} P_0$ will be a sum of terms each one of which can act only by transferring one electron from a singly occupied site, call it i, to another singly occupied site, call it j, so creating a doubly occupied site at j. For the same reasons, $P_0 \mathcal{H} P_\eta$, which multiplies the former term by the left in Eq. (2.52), can only transfer one electron from a doubly occupied site, hence from site j, in the terminology adopted here, to a previously empty site in such a way as to restore the condition expressed by the Gutzwiller projector P_0. The final site can be either the original site i or another site k different from i (with the only restriction that j be n.n. to i and k to j, due to the fact that t_{ij} is assumed to be nonvanishing only between n.n. sites). Site k can be either a site next-nearest neighbor (n.n.n.) to i (e.g. in a square lattice), or another n.n. site (e.g. in a plane triangular lattice). Altogether, the effect of the correction term $-(1/U) P_0 \mathcal{H} P_\eta \mathcal{H} P_0$ in Eq. (2.52) consists of two distinct contributions:

i) A sum of "two-site" terms, corresponding to an electron "jumping" virtually from site i to site j and back. Any such term will be weighted (apart from signs and the overall factor of U^{-1}) by a factor of: $t_{ij} \cdot t_{ji} \equiv |t_{ij}|^2$.

ii) Another sum of "three-site" terms, corresponding to an electron "jumping" from site i to site j and then to site k ($k \neq i$), each term being weighted by a factor of: $t_{ij} \cdot t_{jk}$.

Remark. Terms of the form i) correspond to virtual processes, which are not accompanied by any kind of charge (electron) transfer, while ii) corresponds to an actual charge transfer between different sites. As, in the low-enegy sector (the subspace $\mathcal{E}_0$ spanned by the Gutzwiller projection) this requires the final state to be initially empty, it is quite clear that processes of the type ii) will be completely quenched (as well as the first term in Eq. (2.52)) at exactly half-filling. In such a case, the effective Hamiltonian in the low-energy sector will consist uniquely of the terms of the form i).

We proceed now to evaluate explicitly the effective Hamiltonian of Eq. (2.52). Using Eq. (2.51), we find:

$$P_0 \mathcal{H} P_\eta \mathcal{H} P_0 = P_0 \{\mathcal{H}^{(1)} + \mathcal{H}^{(2)}\} P_0 \tag{2.60}$$

where:

$$\mathcal{H}^{(1)} = \sum_{\langle ij \rangle} \sum_{\sigma\tau} |t_{ij}|^2 (1 - \hat{n}_{i\bar{\sigma}}) c^\dagger_{i\sigma} c_{j\sigma} \hat{n}_{j\bar{\sigma}} \hat{n}_{j\bar{\tau}} c^\dagger_{j\tau} c_{i\tau} (1 - \hat{n}_{i\bar{\tau}}) \tag{2.61}$$

$$\mathcal{H}^{(2)} = \sum_{\langle ijk \rangle} \sum_{\sigma\tau} t_{ij} t_{jk} (1 - \hat{n}_{i\bar{\sigma}}) c^\dagger_{i\sigma} c_{j\sigma} \hat{n}_{j\bar{\sigma}} \hat{n}_{j\bar{\tau}} c^\dagger_{j\tau} c_{k\tau} (1 - \hat{n}_{k\bar{\tau}}) \ . \tag{2.62}$$

In Eq. (2.62), the symbol $\langle ijk \rangle$ denotes a summation in which $i \neq k$, and both i and k are n.n. to j. Equation (2.61) can be recasted in a particularly interesting form. Indeed, rearranging terms, the following result has been obtained by various authors (see in particular Ref. [166]):

$$\mathcal{H}^{(1)} = -\sum_{\langle ij \rangle} 2|t_{ij}|^2 \left\{ \mathbf{S}_i \cdot \mathbf{S}_j - \frac{1}{4} \sum_{\sigma\sigma'} \hat{n}_{i\sigma}(1 - \hat{n}_{i\bar{\sigma}}) \hat{n}_{j\sigma'}(1 - \hat{n}_{j\bar{\sigma}'}) \right\} \tag{2.63}$$

where the spin operators have been defined as in Eq. (2.54).

However, remembering that the above results have to be "sandwiched" between two P_0's, and noting that:

$$\hat{n}_{i\sigma}(1 - \hat{n}_{i\bar{\sigma}}) \equiv n_{i\sigma}(1 - \hat{v}_i) \tag{2.64}$$

one sees easily that:

$$P_0 \sum_{\sigma\sigma'} \hat{n}_{i\sigma}(1 - \hat{n}_{i\bar{\sigma}}) \hat{n}_{j\sigma'}(1 - \hat{n}_{j\bar{\sigma}'}) P_0 \equiv P_0 \hat{n}_i \hat{n}_j P_0 \tag{2.65}$$

so that the final result for the effective Hamiltonian in the low-energy sector is:

$$P_0 \mathcal{H}_{\text{eff}} P_0 = P_0 \tilde{\mathcal{H}}_{\text{eff}} P_0 \tag{2.66}$$

where:

$$\tilde{\mathcal{H}}_{\text{eff}} = T_h + \sum_{\langle ij \rangle} J_{ij} \left\{ \mathbf{S}_i \cdot \mathbf{S}_j - \frac{1}{4} \hat{n}_i \hat{n}_j \right\} - U^{-1} \cdot \mathcal{H}^{(2)} \tag{2.67}$$

and the exchange constants J_{ij} are given by:

$$J_{ij} := 4 \frac{|t_{ij}|^2}{U} \qquad (2.68)$$

Note that the sum in Eq. (2.67) is over unordered pairs, whence the factor of 4 (instead of 2) in Eq. (2.68).

From now on, to shorten the notation and unless explicitly necessary, we will understand that whatever Hamiltonian we write has to be "sandwiched" eventually between two P_0's. The full notation will be resumed whenever necessary.

At exact half-filling, as discussed above, only the second term of Eq. (2.67) survives and, as: $\hat{n}_i \equiv 1 \ \forall i$, we obtain:

$$\tilde{\mathcal{H}}_{\text{eff}}|_{\text{h.f.}} = \sum_{(ij)} J_{ij} \left\{ \mathbf{S}_i \cdot \mathbf{S}_j - \frac{1}{4} \right\}. \qquad (2.69)$$

At half-filling, the Hubbard Hamiltonian is then mapped, to second-order in $|t|/U$, into an antiferromagnetic ($J_{ij} > 0$) Heisenberg Hamiltonian. Away from half-filling, the effective Hamiltonian (again with the restriction provided by the Gutzwiller projection) will be given by the full expression of Eq. (2.67).

3. Some Ideas for RVB

As mentioned in the introduction, the AFM Heisenberg chain was exactly solved by H. Bethe in 1931. The ground state provided by the Bethe Ansatz appears to be dominated by short-range, singlet correlations among spins, and can be described as an admixture of singlet spin-pair configurations. Anderson [17] suggested in 1973 that the same should be true for the two-dimensional isotropic AFM Heisenberg system, and Fazekas and Anderson [68] substantiated this idea with quantitative investigations of a specific model, namely of a system of planar $S = \frac{1}{2}$ spins on a triangular lattice. We give here a sketchy and qualitative account of Fazekas-Anderson's discussion of the triangular lattice. Let us write the Heisenberg spin Hamiltonian as:

$$\mathcal{H} = \mathcal{H}^z + \mathcal{H}^\pm \tag{3.1}$$

where:

$$\mathcal{H}^z = J \sum_{(ij)} S_i^z S_j^z \tag{3.2}$$

is the Ising part of the Hamiltonian, $J > 0$ and:

$$\mathcal{H}^\pm = J \frac{\alpha}{2} \sum_{(ij)} (S_i^+ S_j^- + S_i^- S_j^+) \tag{3.3}$$

α is an anisotropy parameter. $\alpha = 0$ yields the Ising limit of the model, and $\alpha = 1$ corresponds to the physically interesting isotropic case. The lattice can be naturally divided into three sublattices, labeled A, B, C (Fig. 4a), so that spins in sublattice A has only B and C neighbors, and so on. For planar spins, we can take the lattice plane in the, (x-z) plane, say, and set: $S_i^y = 0$. The (classical) ground-state energy corresponds to the spins in sublattices B and C being tilted, in opposite directions, by an angle π-θ with respect to the spin direction in sublattice A (Fig. 4b)). An elementary calculation shows that the best value θ^* of θ is given by:

$$\cos \theta^* = (1 + \alpha)^{-1} \tag{3.4}$$

and the corresponding energy by:

$$E_{cl} = (MS^2J)[-1 - \alpha^2(1 + \alpha)^{-1}] \tag{3.5}$$

with M the total number of lattice sites. The isotropic ($\alpha = 1$) case corresponds to each spin forming an angle of 120° with respect to the spins in the other sublattices. So much for the classical solution.

If (quantum-mechanically) $\mathcal{H}^\pm$ is treated as a perturbation on $\mathcal{H}^z$, the (Ising) ground state of $\mathcal{H}^z$ is a highly frustrated one. For example, the three configurations of 10 spins displayed in Figs. 5a–c have the same number of frustrated bonds, but $\mathcal{H}^\pm$ annihilates c, while it interchanges a and b. A trial ground-state wavefunction can then be constructed

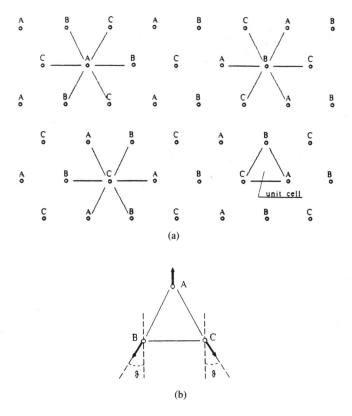

Fig. 4. a) The triangular lattice considered by Fazekas and Anderson. b) General orientation of spins in the elementary cell of a).

by linearly superposing equivalent configurations like (5a) and (5b). A singlet admixture leads to a gain in energy [68]:

$$\frac{\Delta E}{E^{(o)}} = -\frac{\alpha}{2} \qquad (3.6)$$

where: $E^{(o)} = -MJS^2$ is the Ising ground-state energy, to be compared with a $\Delta E/E^{(o)} = O(\alpha^2)$ of the classical result (3.5). A triplet admixture would also lead to (3.6), but with the opposite sign for ΔE, and hence triplet pairing has to be discarded. Singlet bonds like those depicted in Fig. 5d can exist in various spatial configurations (Fig. 6). Fazekas and Anderson's analysis shows that further gains in energy can be obtained by coherently superimposing patterns of singlet bonds in different spatial configurations, i.e. by allowing the singlet bonds to freely "resonate" among different configurations. The most important outcome of this calculation is that the gain in energy appears to survive up to $\alpha = 1$, the (isotropic) case of interest. The picture that emerges for the nature of the ground state is then that of a sort of a (quantum) liquid of singlet bonds which can resonate among different configurations, whence the name of "Resonating Valence Bond" for the proposed ground state.

Anderson and coworkers [18, 21, 38, 41], and also Kivelson *et al.* [117], have proposed that an RVB ground state could be the correct starting point (that is the correct reference ground state) for a theoretical description of the Cu-O planes in the materials which become high-T_c superconductors upon doping. This has led to a renewed interest in the study of the 2D antiferromagnetic Heisenberg Hamiltonian.

There seems to be by now growing evidence coming from numerical studies of finite-size samples [109, 131–133, 154, 162], all on square lattices and with n.n. couplings only, that the 2D antiferromagnetic Heisenberg Hamiltonian does actually have a ground state with well defined antiferromagnetic correlations (a nonvanishing staggered magnetization, for example).

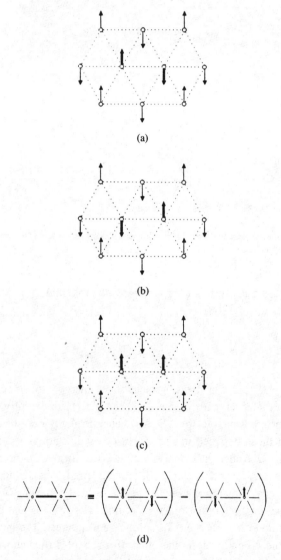

Fig. 5. a) to c): singlet vs. triplet spin configurations on a triangular lattice. d) Pictorial illustration of a Resonating Valence Bond.

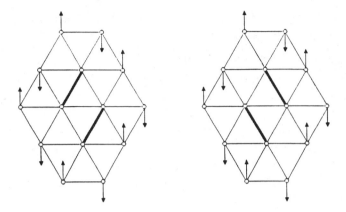

Fig. 6. A pictorial illustration of how two spatially different singlet bonds can resonate in the construction of an RVB state.

Neither of the models considered in the above references exhibit frustration, as it is instead the case for the triangular lattice considered previously. That frustration may favor, if not an RVB, at least a disordered, spin liquid ground state at large values of the spin has been recently argued by Chandra and Doucot [48]. For lower values of the spin, we might expect quantum fluctuations to further favor, in the presence of frustration, disordered ground states.

The relevance of the abovementioned, model-dependent results for the physics of the real materials which exhibit high-T_c superconductivity is also diminished to some extent by the fact that we must in general expect some amount of next-nearest-neighbor coupling (or "superexchange" [14, 16]) to be actually present, thereby introducing frustration. Also, as has been argued in the literature [18, 21, 38, 41, 117], frustration can also be induced by modifications of the exchange integral J_{ij} (cf. Eq. (2.68)) induced by coupling to phonons [117]. Inclusions of these effects may therefore favor again disordered ground states for the 2D Hubbard model, although we must say that, in our opinion, the whole question has not been yet settled in a conclusive and convincing way.

4. RVB Ground States for the Strongly-Coupled Hubbard Hamiltonian at and below Half-Filling. Introduction to Superconductivity

As mentioned in the Introduction, RVB ground states for a half-filled Hubbard Hamiltonian on a 2D square lattice have been studied by Anderson and coworkers [18, 21, 38, 41] and by Kivelson *et al.* [117]. The study of Ref. 117 has certain interesting qualitative aspects, and we begin by briefly reviewing them.

The main features of an RVB ground state and of its elementary excitations are illustrated in Figs. 7 and 8 (taken from Ref. 117). The square lattice can be represented as a bipartite lattice with two interpenetrating ("black" and "white") sublattices. Every black (white) site will have only white (black) n.n.'s. Singlet bonds are then formed between electrons in the two sublattices (Fig. 7a), and the ground state is a coherent superposition of states of this form (Fig. 7b).

As discussed in Sec. 3, there is no frustration in a square lattice to take advantage of in the construction of an RVB state as in the case of the triangular lattice. However, taking into account also the ion-ion interaction (i.e. the phonons), the authors in Ref. 117 argue that: "*Associated with each valence bond is a lattice deformation which increases the hopping energy across that bond . . .*" and that: "*. . . these phonons stabilize the RVB state with respect to the Néel state . . .*", as the latter can occur without any lattice distortion.

To be specific, the Hamiltonian (2.69) is rewritten, including the lattice displacements, in the way that follows.

Assuming, for simplicity, that the hopping integral t_{ij} (Eq. (1.10)) is a function only of $|\mathbf{R}_i - \mathbf{R}_j|$, one can expand in the deviations $\mathbf{u}_i$ and $\mathbf{u}_j$ of $\mathbf{R}_i$ and $\mathbf{R}_j$ from the equilibrium positions. Denoting by $\boldsymbol{\tau}_{ij}$ ($|\boldsymbol{\tau}_{ij}|$ = const. = τ for a square lattice) the equilibrium value of $\mathbf{R}_i - \mathbf{R}_j$, we obtain, to lowest order:

$$t_{ij} = t(|\mathbf{R}_i - \mathbf{R}_j|) \simeq t_0 - \alpha u_{ij} \qquad (4.1)$$

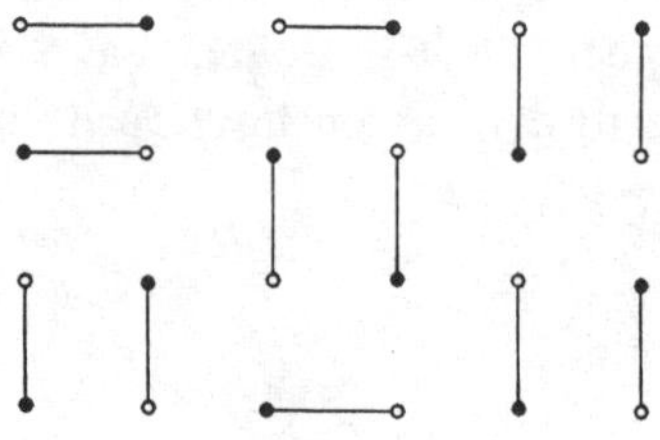

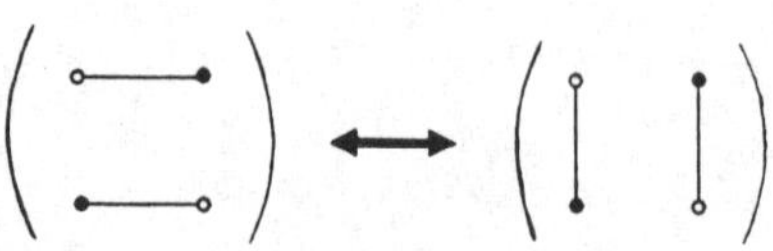

Fig. 7. a) Schematic view of how one can construct a (nearest-neighbour) RVB state on a square 2D lattice. b) Schematic view of how two different singlet bonds can resonate.

where:

$$t_0 = t(\tau) ; \qquad \alpha = -\tau \frac{d}{d\rho} t(\rho) \bigg|_{\rho=\tau} ; \qquad u_{ij} = \tau^{-1}(\mathbf{u}_i - \mathbf{u}_j) \cdot \boldsymbol{\tau}_{ij} . \qquad (4.2)$$

Assuming $t(\rho)$ to be a decreasing function of its argument (at least in the neighborhood of the equilibrium position $\rho = \tau$), α, the electron-phonon coupling constant, will be positive, and t will increase whenever a lattice distortion takes sites i and j closer than the equilibrium distance ($u_{ij} < 0$). The Hamiltonian (2.69) generalizes then into:

$$\mathcal{H} = \sum_{(ij)} \left\{ \frac{4(t_0 - \alpha u_{ij})^2}{U} \mathbf{S}_i \cdot \mathbf{S}_j + \frac{K}{2} |\mathbf{u}_i - \mathbf{u}_j|^2 \right\} \qquad (4.3)$$

where K is the spring constant of the lattice.

The classical Néel state corresponds, for symmetry reasons, to: $\mathbf{u}_i = 0 \ \forall i$, and has an energy of $-2t_0^2/U$ per site. A naive (classical) RVB state can be constructed by taking, say, $|\mathbf{u}_i| = \text{const.} = (u/2)$, $\mathbf{u}_i$ and $\mathbf{u}_j$ pointing towards each other (i.e.: $u_{ij} = -u$) if sites i and j are paired in a singlet bond, and minimizing the resulting (classical) energy with respect to u. It turns out that the RVB state is energetically favoured if $(\alpha^2/KU) > 1/9$. This is of course only an indication that the electron-phonon interaction can indeed stabilize the RVB on a square lattice. It is claimed in Ref. 117 (as well as in [17, 18, 21, 38, 41, 68]) that a proper inclusion of quantum fluctuations further favors the RVB state.

Consider now the effect of breaking a pair, thus creating two dangling bonds, a "black" and a "white" one (Fig. 8). This is clearly an excited state. Each dangling bond has spin-1/2 and charge zero (as, breaking a bond, no charge is added to or subtracted from the system). Each dangling bond can move independently in the lattice, and behaves as an independent elementary excitation, called a "spinon" after Anderson. The situation seems to be similar of that in polyacetylene, where the elementary excitations also appear as neutral solitons [96]. The "black" and "white" spinons appear then as a soliton-antisoliton pair. Attributing to them (arbitrarily) topological charges ± 1, say, the total topological charge enclosed into a large contour cutting no bonds can be counted as the difference between the numbers of enclosed black and white sites, in the manner depicted in Fig. 8. The total topological charge of a state is also a conserved number, as, for the same reasons leading to their creation, solitons and antisolitons can only annihilate in pairs.

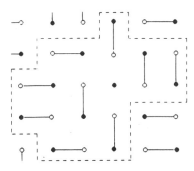

Fig. 8. Creation of a "dangling bond" ("spinon") in an RVB lattice.

Upon doping, the added (charged, spin-1/2) holes can bind to the spinons in a singlet state, thereby forming a charged, spinless soliton called a "holon". By analyzing the statistics of the solitons in the way indicated by Wilczek and Zee [185], i.e. by following the change in the phase of the wavefunction of a pair of identical solitons when the positions of the two are interchanged adiabatically, the authors in Ref. 117 conclude that the spinons are actually fermions, while the charged holons are bosons. This is a delicate point and, as we will see in a later section, many other authors [51, 69, 81, 124, 125, 179, 195] drastically disagree with this conclusion. Also, the fact that the spinon spectrum seems to have a gap in the treatment of Ref. 117 has been questioned by other authors [18, 21, 38, 41]. In particular, Anderson [18] postulated some time before the existence of neutral spin-1/2 elementary excitations over the RVB ground state, but with a gapless excitation spectrum. Be that as it may, an important aspect of the physics of the 2D Hubbard model seems to have been highlighted by this analysis, namely that in the elementary excitation spectrum of a doped Mott-Hubbard insulator there is a separation of the charge and spin degrees of freedom, the former being carried by the holons, the latter by the spinons.

The obvious suggestion that seems to emerge from the analysis of Ref. 117 is that superconductivity can (or could) be attributed to Bose condensation of the holons. However, as discussed in the Introduction, this leads to a prediction for the charge of the order parameter which is half the observed value, and this rules out holon condensation as a mechanism for high-T_c superconductivity.

We now turn to a brief account of the mean-field treatment of the RVB state as given by Anderson and coworkers [18, 21, 38, 41]. Let us start (at or below half-filling) with the Hamiltonian of Eq. (2.67). Employing the identity:

$$(\boldsymbol{\sigma})_{\alpha\beta} \cdot (\boldsymbol{\sigma})_{\mu\nu} = 2\delta_{\alpha\nu}\delta_{\beta\mu} - \delta_{\alpha\beta}\delta_{\mu\nu} \qquad (4.4)$$

one can prove directly that:

$$\mathbf{S}_i \cdot \mathbf{S}_j - \frac{1}{4}\hat{n}_i\hat{n}_j \equiv -b_{ij}^\dagger b_{ij} \qquad (4.5)$$

where

$$b_{ij}^\dagger =: \frac{1}{\sqrt{2}}[c_{i\uparrow}^\dagger c_{j\downarrow}^\dagger - c_{i\downarrow}^\dagger c_{j\uparrow}^\dagger] \equiv b_{ji}^\dagger. \qquad (4.6)$$

The operator $b_{ij}^\dagger$ creates a pair of electrons in a singlet state on sites i and j. We will be interested only in the case when i and j are n.n. sites. Unfortunately, the algebra of the operators (4.6) is rather complicated, as the following commutation relations show:

$$[\![b_{ij}, b_{jk}^\dagger]\!] = \delta_{ik}\left(1 - \frac{\hat{n}_j}{2}\right) - \frac{1}{2}\sum_\sigma c_{k\sigma}^\dagger c_{i\sigma} \qquad (4.7)$$

$((i, j)$ and (j, k) being n.n.) while all the other commutators vanish [104]. In particular:

$$[\![b_{ij}, b_{ij}^\dagger]\!] = 1 - \frac{\hat{n}_i + \hat{n}_j}{2} \tag{4.8}$$

for $i = k$, and:

$$[\![b_{ij}, b_{jk}^\dagger]\!] = -\frac{1}{2} \sum_\sigma c_{k\sigma}^\dagger c_{i\sigma} \tag{4.9}$$

for $i \neq k$.

Even at half-filling, when (4.8) vanishes, the commutators are noncanonical, and the Heisenberg Hamiltonian:

$$\mathcal{H}|_{\text{h.f.}} = J \sum_{(ij)} \left[\mathbf{S}_i \cdot \mathbf{S}_j - \frac{1}{4} \right] \equiv -J \sum_{(ij)} b_{ij}^\dagger b_{ij} \tag{4.10}$$

remains highly nontrivial. In other words, the b's and $b^\dagger$'s act as dynamical link variables, and the noncanonicity of the commutation relations expresses the fact that the link dynamics is highly nontrivial.

As discussed in Sec. 2, the effective Hamiltonian (2.67) is supposed to be restricted to the subspace $\mathcal{E}_0$ of the full Fock space which is spanned by the Gutzwiller projection. Such a restriction is rather difficult to handle properly at every stage of the calculations, and Anderson and coworkers resorted to the assumption that its main effect should be that of reducing the hopping amplitude, in the effective hopping terms T_h and $\mathcal{H}^{(2)}$, below its nominal value t_{ij} to: $\delta \cdot t_{ij}$, where δ is the doping fraction ($\delta = 1 - (N/M)$ in the terminology of Sec. 2). With the further approximation of setting t_{ij} equal to a constant t, i.e. altogether:

$$t_{ij} \simeq t \cdot \delta \tag{4.11}$$

and neglecting the three-site hopping term which, with the approximation (4.11), turns out to be of order δ^2, and hence negligible for $\delta \ll 1$, the effective Hamiltonian reduces to:

$$\tilde{\mathcal{H}}_{\text{eff}} = -t\delta \sum_{(ij)\sigma} (c_{i\sigma}^\dagger c_{j\sigma} + \text{h.c.}) - J \sum_{(ij)} b_{ij}^\dagger b_{ij} . \tag{4.12}$$

Anderson and coworkers have developed a mean-field theory of RVB starting from the Hamiltonian (4.12), letting it act on the full Hilbert space of states, and finally projecting the resulting ground state (as well as the excited states) onto the appropriate subspace. The approximations leading to this mean field theory have not yet received a precise justification, though they find some legitimacy in the fact that, when applied to the one-dimensional Hubbard chain, the theory gives results [38, 41] in good agreement with the exact, Bethe-Ansatz solution [44].

The negative sign in the second term in (4.12) seems to suggest that electrons have a tendency to bind into n.n. singlet pairs, and that an appropriate variational ground state should include correlations leading to a nonvanishing value of:

$$\Delta_{ij} =: \sqrt{2}\langle b_{ij}\rangle \tag{4.13}$$

(the factor of $\sqrt{2}$ being included only for convenience, see Eq. (4.6)). Therefore, this situation is very close to that occurring in the BCS treatment of superconductivity, and suggests a Hartree-Fock BCS-like [156] factorization of the Hamiltonian (4.12).

Another expectation value which should be factored out in a Hartree-Fock approximation is suggested by the nature of the hopping term, and is:

$$p_{ij} =: \langle c_{i\sigma}^\dagger c_{j\sigma}\rangle \; ; \qquad i,j = \text{n.n.} \; . \tag{4.14}$$

Let us approximate, following Refs. [21, 38, 41]:

$$\Delta_{ij} \simeq \text{const.} = \Delta \; ; \qquad p_{ij} \simeq \text{const.} = p \; ; \qquad p \text{ and } \Delta \text{ real} \; . \tag{4.15}$$

(A more general treatment will be reviewed in the next section. In the language to be established there, the first of Eqs. (4.15) corresponds to the so-called "uniform phase".)

It is useful at this point to revert from Wannier to band states using Eq. (1.15). After some long but straightforward algebra, factoring out the expectation values (4.13)–(4.14), and including also a chemical potential term [43, 156], with the chemical potential acting, as usual [122], as a Lagrangian multiplier fixing the average number of particles, the effective Hartree-Fock Hamiltonian becomes:

$$\mathcal{H}_{\text{eff}} \simeq \mathcal{H}_{\text{HF}} = \sum_{\mathbf{k}\sigma}(\epsilon_{\mathbf{k}} - \mu)c_{\mathbf{k}\sigma}^\dagger c_{\mathbf{k}\sigma} - J\sum_{\mathbf{k}}[\Delta\gamma_{\mathbf{k}}c_{\mathbf{k}\uparrow}^\dagger c_{-\mathbf{k}\downarrow}^\dagger + \text{h.c.}] - M[\Delta^2 + p^2] \tag{4.16}$$

where the sums over $\mathbf{k}$ run over the first Brillouin zone, μ is the chemical potential and, for a square 2D lattice with lattice constant a:

$$\gamma_{\mathbf{k}} = \cos(k_x a) + \cos(k_y a) \; ; \qquad \epsilon_{\mathbf{k}} = -(2t\delta + pJ)\gamma_{\mathbf{k}} \; . \tag{4.17}$$

Note that: $\gamma_{\mathbf{k}} = 0$ along the "surface" in the Brillouin zone corresponding to exact half-filling (cf. Fig. 2).

In the standard language of BCS theory [43, 156], one defines a canonical transformation to "quasiparticle" operators:

$$\begin{cases} \alpha_{\mathbf{k}} = u_{\mathbf{k}}c_{\mathbf{k}\uparrow} - v_{\mathbf{k}}c_{-\mathbf{k}\downarrow}^\dagger \\ \beta_{\mathbf{k}} = v_{\mathbf{k}}c_{\mathbf{k}\uparrow}^\dagger + u_{\mathbf{k}}c_{-\mathbf{k}\downarrow} \end{cases} \tag{4.18}$$

with: $|u_{\mathbf{k}}|^2 + |v_{\mathbf{k}}|^2 = 1 \; \forall \mathbf{k}$. At $T = 0$ one also works with a variational ground state of the form:

$$|\psi\rangle = \prod_{\mathbf{k}} (u_{\mathbf{k}} + v_{\mathbf{k}} b_{\mathbf{k}}^\dagger)|0\rangle \tag{4.19}$$

where $|0\rangle$ is the Fock vacuum, and:

$$b_{\mathbf{k}}^\dagger =: c_{\mathbf{k}\uparrow}^\dagger c_{-\mathbf{k}\downarrow}^\dagger . \tag{4.20}$$

The state $|\psi\rangle$ turns out to be the Fock vacuum of the quasiparticle operators.
The BCS theory is defined in terms of the "anomalous" expectation values:

$$\langle \psi | b_{\mathbf{k}} | \psi \rangle =: \langle b_{\mathbf{k}} \rangle \equiv \langle c_{-\mathbf{k}\downarrow} c_{\mathbf{k}\uparrow} \rangle = u_{\mathbf{k}} \bar{v}_{\mathbf{k}} \tag{4.21}$$

and the order-parameter Δ is given by:

$$\Delta = \frac{1}{M} \sum_{\mathbf{k}} \gamma_{\mathbf{k}} u_{\mathbf{k}} \bar{v}_{\mathbf{k}} . \tag{4.22}$$

The values of the coefficients of the canonical transformation are determined, at $T = 0$, by the requirement that (4.18) diagonalizes (4.16). More generally, at $T \neq 0$, they are determined by the requirement [135, 165] that the free-energy:

$$F =: -\beta^{-1} \ln\{\mathrm{Tr}\exp[-\beta \mathcal{H}_{\mathrm{HF}}]\} \; ; \quad \beta = (kT)^{-1} \tag{4.23}$$

be at least a local minimum of $u_{\mathbf{k}}$, $v_{\mathbf{k}}$ subject to:

$$|u_{\mathbf{k}}|^2 + |v_{\mathbf{k}}|^2 = 1 \; \forall \mathbf{k} . \tag{4.24}$$

Finally, the chemical potential has to be determined in such a way that the average number of particles is equal to a fixed number $N \leq M$, i.e.:

$$N = M(1 - \delta) . \tag{4.25}$$

The minimization procedure is standard, and leads to the equations:

$$\frac{1}{M} \sum_{\mathbf{k}} \mathrm{th}(\beta E_{\mathbf{k}}) \cdot \left(\frac{\gamma_{\mathbf{k}}^2}{2E_{\mathbf{k}}} \right) = \frac{1}{J} \tag{4.26}$$

for Δ (after discarding the trivial solution $\Delta = 0$),

$$p = -\frac{1}{2M} \sum_{\mathbf{k}} \gamma_{\mathbf{k}} \, \mathrm{th}\left(\frac{\beta E_{\mathbf{k}}}{2}\right) \cdot \left[\frac{(\epsilon_{\mathbf{k}} - \mu)}{2E_{\mathbf{k}}}\right] \tag{4.27}$$

for p, and:

$$\frac{1}{M} \sum_{\mathbf{k}} \mathrm{th}\left(\frac{\beta E_{\mathbf{k}}}{2}\right) \cdot \left[\frac{(\epsilon_{\mathbf{k}} - \mu)}{E_{\mathbf{k}}}\right] = \delta \tag{4.28}$$

as the equation for the chemical potential, where the quasiparticle energies are given by:

$$E_{\mathbf{k}} = \{(\epsilon_{\mathbf{k}} - \mu)^2 + J^2\Delta^2\gamma_{\mathbf{k}}^2\}^{1/2} \ . \tag{4.29}$$

Let us recall that, according to the standard formulation of the BCS theory [156]:

$$|u_{\mathbf{k}}|^2 - |v_{\mathbf{k}}|^2 = \frac{(\epsilon_{\mathbf{k}} - \mu)}{E_{\mathbf{k}}} \tag{4.30}$$

which, together with (4.24), implies:

$$|u_{\mathbf{k}}|^2 = \frac{1}{2}\left[1 + \frac{(\epsilon_{\mathbf{k}} - \mu)}{E_{\mathbf{k}}}\right] ; \quad |v_{\mathbf{k}}|^2 = \frac{1}{2}\left[1 - \frac{(\epsilon_{\mathbf{k}} - \mu)}{E_{\mathbf{k}}}\right] . \tag{4.31}$$

It is also known that a BCS-type ground state like (4.19) breaks the (global) U(1) gauge invariance of the theory, expressed by the transformation:

$$c_{\mathbf{k}\sigma}^\dagger \to \exp(i\phi)c_{\mathbf{k}\sigma}^\dagger ; \quad \phi \in \mathbb{R} \bmod (2\pi) \tag{4.32}$$

down to a $\mathbb{Z}_2$ invariance [174]. The invariance of the dynamics under the automorphism (4.32) implies that the self-consistency equations of the BCS theory determine only the relative phase of $u_{\mathbf{k}}$ and $v_{\mathbf{k}}$, the product $u_{\mathbf{k}} \cdot \bar{v}_{\mathbf{k}}$ remaining undetermined by an overall (**k**-independent) phase. All this material, which will be of some use shortly below, will be reviewed in Appendix A.

Let us consider first the $T = 0$ case at exactly half-filling: $\delta = 0$. Then: $\epsilon_{\mathbf{k}} = pJ\gamma_{\mathbf{k}}$, and Eqs. (4.27)–(4.28) are trivially satisfied by: $p = \mu = 0$ ($\mu = 0$ is also the $T = 0$ chemical potential when $\delta = J = 0$). As $\epsilon_{\mathbf{k}} = 0$ as well, it follows from Eq. (4.31) that:

$$|u_{\mathbf{k}}|^2 = |v_{\mathbf{k}}|^2 = \frac{1}{2} \tag{4.33}$$

and Eq. (4.26) can be solved at once, yielding:

$$\Delta = \frac{1}{2M}\sum_{\mathbf{k}} |\gamma_{\mathbf{k}}| \ . \tag{4.34}$$

Comparison with (4.22) implies that the u's and v's can be chosen to be real, and:

$$u_{\mathbf{k}} \cdot v_{\mathbf{k}} = \frac{1}{2}\mathrm{sgn}(\gamma_{\mathbf{k}}) \ . \tag{4.35}$$

Therefore, we can choose the solution [21, 38, 41]:

$$\begin{cases} u_{\mathbf{k}} = v_{\mathbf{k}} = \dfrac{1}{\sqrt{2}} & \text{if } \gamma_{\mathbf{k}} > 0 \\[1em] u_{\mathbf{k}} = -v_{\mathbf{k}} = \dfrac{1}{\sqrt{2}} & \text{if } \gamma_{\mathbf{k}} < 0 \end{cases} \tag{4.36}$$

and the quasiparticle energies become:

$$E_{\mathbf{k}} = J\Delta|\gamma_{\mathbf{k}}| . \tag{4.37}$$

Explicitly, the quasiparticle operators will be given by:

$$\begin{cases} \alpha_{\mathbf{k}} = \dfrac{1}{\sqrt{2}} [c_{\mathbf{k}\uparrow} - \mathrm{sgn}(\gamma_{\mathbf{k}}) c^{\dagger}_{-\mathbf{k}\downarrow}] \\ \beta_{\mathbf{k}} = \dfrac{1}{\sqrt{2}} [\mathrm{sgn}(\gamma_{\mathbf{k}}) c^{\dagger}_{\mathbf{k}\uparrow} + c_{-\mathbf{k}\downarrow}] \end{cases} \tag{4.38}$$

The quasiparticle energy (4.37) vanishes at a "pseudo-Fermi surface" (PFS) defined by the condition $\gamma_{\mathbf{k}} = 0$. The PFS is, in this case, again the inscribed square of Fig. 2. Note that the PFS nests, but, because of the low dimensionality of the system no real transition to a static spin-density-wave (SDW) state is actually permitted.

The BCS ground state (4.19) can be rewritten in general as [156]:

$$|\psi\rangle = A \exp\left[\sum_{\mathbf{k}} g(\mathbf{k}) b^{\dagger}_{\mathbf{k}}\right] |0\rangle \tag{4.39}$$

with A a normalization factor, and:

$$g(\mathbf{k}) =: \left(\dfrac{v_{\mathbf{k}}}{u_{\mathbf{k}}}\right) . \tag{4.40}$$

In our case, $g(\mathbf{k}) \equiv \mathrm{sgn}(\gamma_{\mathbf{k}})$ changes sign [18] at the PFS, and satisfies:

$$\sum_{\mathbf{k}} g(\mathbf{k}) = 0 . \tag{4.41}$$

The meaning of the condition (4.41) can be best seen by going back to the Wannier representation. Then:

$$b^{\dagger} =: \sum_{\mathbf{k}} g(\mathbf{k}) b^{\dagger}_{\mathbf{k}} = \frac{1}{M} \sum_{ij} \tilde{g}(\mathbf{R}_i - \mathbf{R}_j) c^{\dagger}_{i\uparrow} c^{\dagger}_{j\downarrow} \tag{4.42}$$

where:

$$\tilde{g}(\mathbf{R}) =: \sum_{\mathbf{k}} \exp[i\mathbf{k} \cdot \mathbf{R}] g(\mathbf{k}) . \tag{4.43}$$

It follows that the operator (4.42) creates a coherent superposition of singlet bonds with variable lengths ($|\mathbf{R}_i - \mathbf{R}_j|$), each weighted by a factor of $\tilde{g}(|\mathbf{R}_i - \mathbf{R}_j|)$, and (4.41) simply means that $\tilde{g}(0)$ vanishes, i.e. that $b^{\dagger}|0\rangle$ contains no doubly-occupied sites.

However, this fails to be true for higher powers of $b^\dagger$ acting on the (true) Fock vacuum in the expansion of the r.h.s. of (4.39), and one has then to proceed to a careful projection of the BCS state. This is done in two steps: first projecting out of the BCS state the component with the appropriate number N of particles (i.e. with $N/2$ singlet pairs), and then applying the Gutzwiller projection to avoid double occupancies. The properties of projected BCS states are reviewed in Appendix A to the present paper, and we will refer to it for details.

Denoting by P_N the projection onto the N-particle subspace of the Fock space, the N-particle component of (4.39) will be:

$$P_N|\psi\rangle = \frac{A}{n!}(b^\dagger)^{N/2}|0\rangle \qquad (4.44)$$

and can also be obtained (see Appendix A) by averaging over phases in the following manner:

$$P_N|\psi\rangle = A \int_0^{2\pi} \frac{d\phi}{2\pi} \exp[-iN\phi] \exp\{b^\dagger \exp(2i\phi)\}|0\rangle \qquad (4.45)$$

by exploiting the action of the (spontaneously broken) U(1) symmetry on the ground state. As discussed in Appendix A, averaging over phases as in (4.45) restores the gauge invariance of the theory and, as a consequence of this, there is no true off-diagonal long-range order (ODLRO [190]) in the (projected) N-particle state.

The final state which is a candidate to represent the RVB ground state will be, according to what has been said above, and apart from normalization:

$$|\text{RVB}\rangle = P_0 P_N |\psi\rangle \qquad (4.46)$$

with P_0 the Gutzwiller projection.

The elementary excitations over the projected state $P_N|\psi\rangle$ are obtained, as discussed in Appendix A, by applying pairs of quasiparticle creation operators to the BCS ground state, that is they are of the form: $P_N \alpha_{\mathbf{k}}^\dagger \alpha_{\mathbf{k}'}^\dagger |\psi\rangle$, $P_N \beta_{\mathbf{k}} \beta_{\mathbf{k}'} |\psi\rangle$, and so on. After projecting with the Gutzwiller projection, those which survive become elementary excitations over the RVB ground state.

The authors in Ref. [41] have discussed the elementary excitations created by the application of $\alpha_{\mathbf{k}}^\dagger \alpha_{\mathbf{k}'}^\dagger$ to the ground state (see Eq. (A31)).

They claim that, when $\mathbf{k}$ and $\mathbf{k}'$ are on the same side of the PFS, $\alpha_{\mathbf{k}}^\dagger \alpha_{\mathbf{k}'}^\dagger$ creates a charged excitation which does not survive the Gutzwiller projection (that is, it is not an excitation of the RVB state). On the contrary, when $\mathbf{k}$ and $\mathbf{k}'$ are on opposite sides of the PFS, the excitations are gapless in the insulating state (cf. Eq. (4.37)), and resemble [18] the elementary excitations of a Fermi liquid. Being gapless, this excitation spectrum gives rise to a linear term in the specific heat and to a Pauli-like susceptibility, in qualitative agreement with the experiment. These excitations across the PFS have been identified by the above authors with the spinons of Kivelson, Rokhsar and Sethna [117].

We turn now to the effect of doping (that is to the case $\delta \neq 0$), restricting again, for simplicity, to the zero-temperature limit. As already mentioned in the Introduction, as soon as holes are introduced, the system metallizes, and a gap opens up at the (pseudo) Fermi surface, giving rise to superconductivity.

In order to obtain a quantitative idea of how this happens, let us restrict ourselves to $\delta \ll 1$, setting, as in Refs. [21, 38, 41], $p = 0$ and neglecting, for the time being, the dependence of the gap parameter Δ upon the doping fraction δ.

At $T = 0$, the equation for the chemical potential, Eq. (4.28), becomes:

$$\frac{1}{M} \sum_{\mathbf{k}} \frac{\epsilon_{\mathbf{k}} - \mu}{E_{\mathbf{k}}} = \delta \qquad (4.47)$$

which can also be rewritten (cf. Eq. (4.29)) as:

$$\delta = -\frac{\partial}{\partial \mu} \left\{ \frac{1}{M} \sum_{\mathbf{k}} E_{\mathbf{k}} \right\}. \qquad (4.48)$$

When $T > 0$, a similar equation holds, with $E_{\mathbf{k}}$ substituted by: $2\beta^{-1} \ln \mathrm{ch}(\beta E_{\mathbf{k}}/2)$ in the r.h.s. of (4.48). The form (4.48) of the equation for the chemical potential can be useful for practical computations.

Under the simplifying assumptions listed above, differentiating Eq. (4.47) with respect to δ, we obtain:

$$1 = -\frac{\partial \mu}{\partial \delta} \cdot \frac{1}{M} \sum_{\mathbf{k}} \left\{ \frac{J^2 \Delta^2 \gamma_{\mathbf{k}}^2}{E_{\mathbf{k}}^3} \right\}. \qquad (4.49)$$

Hence:

$$\frac{\partial \mu}{\partial \delta} < 0. \qquad (4.50)$$

Introducing holes ($\delta > 0$) has then the effect of driving the chemical potential below the value ($\mu = 0$) it has in the insulating state ($\delta = 0$). μ itself will be (negative and) proportional to δ for small values of δ. It is easily seen from Eq. (4.29) that, as a consequence, $E_{\mathbf{k}}$ will indeed develop a gap of width: $E_{\mathrm{gap}} \propto |\mu|$ (and hence also $\propto \delta$) at the PFS. Actually, the authors in Ref. [41] find:

$$E_{\mathrm{gap}} = \frac{3\delta \Delta J/2}{\sqrt{1 + \left(\frac{3\Delta J}{2W}\right)^2}}. \qquad (4.51)$$

They state that: "For some choice of the parameters, $2E_{\mathrm{gap}}/T_c$ is close to the BCS value". The "BCS value" is the rather famous value of $\simeq 3.52$ for the ratio of twice the zero-temperature gap (which is also the true energy gap for excitations in systems with a fixed number of particles (see App. A)) to the transition temperature one finds in the standard BCS theory [156].

The same authors have worked out (numerically) the superconducting transition temperature T_c as a function of the doping fraction δ (Fig. 1 of Ref. [41], and Fig. 3 of Ref. [21]). Their mean field theory has been also critically discussed in Ref. [104], where it has been pointed out that there is a severe mismatch between the predicted dependence of T_c on δ and the existing experimental data, mainly in the low doping ($\delta \lesssim 0.2$) part of the phase diagram.

As a final comment to this section, it seems to us that the approach presented in Refs. [21, 38, 41] is full of interesting physical ideas and proposals, and that it substantiates to some extent the ideas that Anderson had put forward, with remarkable hindsight, some time before [18]. Satisfactory proofs of many of the arguments however are still lacking.

In the next section, adopting also a functional-integral point of view, we will review how this whole set of ideas has been elaborated further, to yield further insights into the nature of the RVB ground state of the 2D Hubbard model.

5. Other Mean-Field Theories for the Hubbard-Heisenberg Hamiltonian on a 2D Square Lattice: Non-Uniform and Flux Phases

What we have reviewed in the last part of Sec. 4 is essentially a rather simplified mean field theory which assumed a "uniform" order parameter, i.e. a link variable Δ with the same value for all n.n. link directions (and zero otherwise). We want here to discuss other and more general mean-field theories that have been proposed for the 2D Hubbard model at and below half-filling. We will start with a more accurate discussion of the mean field theory arising from the Hartree-Fock factorization proposed in Refs. [21, 38, 41], thus making also contact with the results obtained by Kotliar [120]. Details on most of the material we will be using in this section can be found in Ref. [66].

Let us start again with the Hamiltonian of Eq. (4.12), and with the Hartree-Fock factorizations (4.13)–(4.14). We will keep the approximation of setting:

$$p_{ij} = \text{const.} = p \; ; \qquad p \text{ real} \tag{5.1}$$

but will make no *a priori* assumptions on Δ_{ij}, except that $\Delta_{ij} \neq 0$ only when (i, j) are nearest neighbors, and that the Δ_{ij}'s are translationally invariant.

Including again the chemical potential, the effective Hamiltonian becomes then:

$$\mathcal{H}_{\text{eff}} = \sum_{(ij)\sigma} (\epsilon c_{i\sigma}^\dagger c_{j\sigma} + \text{h.c.}) - \frac{J}{2} \sum_{(ij)} \{\Delta_{ij} b_{ij}^\dagger + \text{h.c.}\}$$

$$- \frac{J}{2} \sum_{(ij)} \{|\Delta_{ij}|^2 + 2p^2\} - \mu \sum_{i\sigma} c_{i\sigma}^\dagger c_{i\sigma} \tag{5.2}$$

where:

$$\epsilon =: -t\delta - \frac{J}{2} p \; . \tag{5.3}$$

We next define the relevant (two-point) imaginary-time Green functions [2] as:

$$\mathcal{G}_\sigma^{ij}(\tau) =: -\langle T_\tau \{c_{i\sigma}(\tau) c_{j\sigma}^\dagger(0)\}\rangle \tag{5.4a}$$

$$\mathcal{F}_\alpha^{ij\dagger}(\tau) =: -\langle T_\tau \{c_{i\alpha}^\dagger(\tau) c_{j\bar\alpha}^\dagger(0)\}\rangle \tag{5.4b}$$

where $|\tau| \leq \beta\hbar$, σ and α are spin indices, and:

$$c_{i\sigma}(\tau) =: \exp\left[\frac{\mathcal{H}_{\text{eff}}\tau}{\hbar}\right] c_{i\sigma} \exp\left[-\frac{\mathcal{H}_{\text{eff}}\tau}{\hbar}\right] . \tag{5.5}$$

As is well-known [2], the (Kubo-Martin-Schwinger) boundary conditions obeyed by the Green functions in the imaginary-time domain yield the following Fourier representation(s):

$$\mathcal{G}_\sigma^{ij}(\tau) = \frac{1}{\beta\hbar} \sum_n \mathcal{G}_\sigma^{ij}(\omega_n) \exp[-i\omega_n\tau] \tag{5.6}$$

where:

$$\omega_n = \frac{(2n+1)\pi}{\beta\hbar}; \quad n \in \mathbb{Z}; \tag{5.7}$$

with a similar representation holding for the second Green function (5.4b), i.e.:

$$\mathcal{F}_\alpha^{ij\dagger}(\tau) = \frac{1}{\beta\hbar} \sum_n \mathcal{F}_\alpha^{ij\dagger}(\omega_n) \exp(-i\omega_n\tau). \tag{5.8}$$

The equations of motion for the Green functions are of the standard Gorkov form [2], and can be converted into a closed algebraic system by Fourier transforming also with respect to the dependence on the space coordinates.

Let us define:

$$\mathcal{U}(\mathbf{k}) =: \frac{1}{2} \sum_{\boldsymbol{\tau}_{ij}} \Delta_{ij} \exp[i\mathbf{k} \cdot \boldsymbol{\tau}_{ij}] \tag{5.9}$$

where:

$$\boldsymbol{\tau}_{ij} \equiv \mathbf{R}_i - \mathbf{R}_j \tag{5.10}$$

is the vector connecting any site i to its nearest neighbors.

The symmetry relation (see Sec. 4) $b_{ij} = b_{ji}$ implies of course:

$$\Delta_{ij} = \Delta_{ji} \tag{5.11}$$

i.e. that the link variables have no "directional" character in this approach. The additional assumption of translational invariance on the square lattice implies that Δ_{ij} is entirely characterized by its values along the two coordinate axes of the lattice, Δ_x and Δ_y, and hence:

$$\mathcal{U}(\mathbf{k}) = \Delta_x \cos(k_x a) + \Delta_y \cos(k_y a). \tag{5.12}$$

Proceeding as in Sec. 4, we define the single-particle energies as:

$$\epsilon(\mathbf{k}) =: -(2t\delta + pJ) \cdot \gamma(\mathbf{k}) \tag{5.13}$$

(cf. Eq. (4.17)), and the quasi-particle energies as:

$$E(\mathbf{k}) =: \sqrt{(\epsilon(\mathbf{k}) - \mu)^2 + J^2|\mathcal{U}(\mathbf{k})|^2}. \tag{5.14}$$

Explicitly, denoting by $\mathcal{F}_\alpha^\dagger(\mathbf{k}, \omega_n)$ and $\mathcal{G}_\sigma(\mathbf{k}, \omega_n)$ the (full) Fourier transforms of the Green functions, we find:

$$\mathcal{F}_\downarrow^\dagger(\mathbf{k}, \omega_n) = -\mathcal{F}_\uparrow^\dagger(\mathbf{k}, \omega_n) = \frac{\hbar J \mathcal{U}^*(\mathbf{k})}{\hbar^2 \omega_n^2 + (\epsilon(\mathbf{k}) - \mu)^2 + J^2 |\mathcal{U}(\mathbf{k})|^2} \quad (5.15)$$

and a spin-independent solution for $\mathcal{G}_\uparrow(\mathbf{k}, \omega_n) \equiv \mathcal{G}_\downarrow(\mathbf{k}, \omega_n)$, which will however be of no concern for us (for more details, see Ref. [66]).

The Hartree-Fock self-consistency condition is expressed by:

$$\Delta_{ij}^* = -\lim_{\tau \to +0} [\mathcal{F}_\uparrow^{ij\dagger}(\tau) - \mathcal{F}_\downarrow^{ij\dagger}(\tau)] . \quad (5.16)$$

Note that, in Eq. (5.16), the sites i and j (by the very definition of the order parameter Δ_{ij}) are to be understood as n.n. sites.

Explicitly, we find, after some long but straightforward and standard [2] algebra:

$$\Delta_{ij} = \frac{J}{M} \sum_\mathbf{k} \exp[i\mathbf{k} \cdot \boldsymbol{\tau}_{ij}] \cdot \frac{\mathcal{U}(\mathbf{k})}{E(\mathbf{k})} \tanh\left[\frac{\beta E(\mathbf{k})}{2}\right] \quad (5.17)$$

where, as before, the $\mathbf{k}$-sum runs over the first Brillouin zone.

Remark: If both sides of Eq. (5.17) are Fourier-transformed using Eq. (5.9), we obtain exactly the same self-consistency equation as obtained by Kotliar in Ref. [120], which generalizes the self-consistency condition of Refs. [21, 38, 41].

Taking now $\boldsymbol{\tau}_{ij}$ in the two independent directions x and y (cf. Eq. (5.10)), we obtain from (5.17) the coupled system of (nonlinear) self-consistency equations (recall that the lattice parameter has been set equal to a):

$$\Delta_x = \frac{J}{M} \sum_\mathbf{k} \exp[ik_x a] \frac{\Delta_x \cos(k_x a) + \Delta_y \cos(k_y a)}{E(\mathbf{k})} \tanh\left(\frac{\beta E(\mathbf{k})}{2}\right) \quad (5.18a)$$

$$\Delta_y = \frac{J}{M} \sum_\mathbf{k} \exp[ik_y a] \frac{\Delta_x \cos(k_x a) + \Delta_y \cos(k_y a)}{E(\mathbf{k})} \tanh\left(\frac{\beta E(\mathbf{k})}{2}\right) . \quad (5.18b)$$

The equation for the chemical potential is the same as in Sec. 4, and will not be rediscussed here. Also, as p acts simply to renormalize the single-particle energies, we will take $p = 0$ in what follows, as done in Refs. [21, 41] and [120].

For any given solution of Eq. (5.18), another one can be generated with the interchange:

$$\Delta_x \Leftrightarrow \Delta_y . \quad (5.19)$$

(This can be seen most easily by interchanging the "dummy" variables k_x and k_y in the equations.) Equation (5.19) is of course only a direct consequence of the cubic symmetry of the problem.

Let us fix the phase of the gap parameter in such a way that Δ_x is real. A class of solutions to Eqs. (5.18) that has been investigated in Ref. [120] corresponds to the two Δ's being equal in magnitude, but with different relative phases. Setting then:

$$\Delta_x = e^{-i\phi}\Delta_y = \Delta ; \quad \Delta \text{ real} \quad (5.20)$$

we have:

$$E(\mathbf{k}) = \sqrt{(\epsilon(\mathbf{k}) - \mu)^2 + J^2\Delta^2(\cos^2(k_x a) + \cos^2(k_y a) + 2\cos\phi \cdot \cos(k_x a)\cos(k_y a))}. \quad (5.21)$$

Note that: $E(\mathbf{k}) \equiv E(k_x, k_y)$ in Eq. (5.21) exhibits full cubic symmetry, namely it is an even function of both its arguments, and it is symmetric under the interchange $k_x \Leftrightarrow k_y$.

Using Eq. (5.20), (5.18a) becomes (for $\Delta \neq 0$):

$$1 = \frac{J}{M} \sum_{\mathbf{k}} \exp[ik_x a] \frac{\cos(k_x a) + e^{i\phi}\cos(k_y a)}{E(\mathbf{k})} \tanh\left(\frac{\beta E(\mathbf{k})}{2}\right) \quad (5.22)$$

while, interchanging k_x and k_y in the summand, (5.18b) turns into the same equation, but with $\phi \Leftrightarrow -\phi$. For symmetry reasons (sending $k_x \to -k_x$ in the summand), Eq. (5.22) further reduces to:

$$1 = \frac{J}{M} \sum_{\mathbf{k}} \cos(k_x a) \frac{\cos(k_x a) + e^{i\phi}\cos(k_y a)}{E(\mathbf{k})} \tanh\left(\frac{\beta E(\mathbf{k})}{2}\right) \quad (5.23)$$

together with the parent equation with $\phi \to -\phi$. Dividing into real and imaginary parts, we find that, unless the condition

$$\frac{1}{M} \sum_{\mathbf{k}} \frac{\cos(k_x a)\cos(k_y a)}{E(\mathbf{k})} \tanh\left(\frac{\beta E(\mathbf{k})}{2}\right) = 0 \quad (5.24)$$

holds, ϕ is forced to be zero or π. This hence gives rise to the following solutions, if we do not insist on (5.24):

i) $\phi = 0$ ($\Delta_x = \Delta_y$, "s-wave" solution) yields:

$$\mathcal{U}(\mathbf{k}) = \Delta \cdot \gamma(\mathbf{k}) \quad (5.25)$$

and a quasi-particle spectrum:

$$E(\mathbf{k}) = \sqrt{(\epsilon(\mathbf{k}) - \mu)^2 + J^2\Delta^2\gamma(\mathbf{k})^2} . \quad (5.26)$$

By the same symmetry arguments as employed before, Eq. (5.23) can be shown to coincide with (4.26). This "uniform" or "s-wave" solution is then the one given in Refs. [21, 41], and already discussed in Sec. 4.

ii) $\phi = \pi$ ($\Delta_x = -\Delta_y$; "d-wave" solution) yields:

$$\mathcal{U}(\mathbf{k}) = \Delta(\cos(k_x a) - \cos(k_y a)) \tag{5.27}$$

and the quasi-particle spectrum:

$$E(\mathbf{k}) = \sqrt{(\epsilon(\mathbf{k}) - \mu)^2 + J^2 \Delta^2 (\cos(k_x a) - \cos(k_y a))^2} . \tag{5.28}$$

At exact half-filling ($\delta = 0$, implying (see Sec. 4) $\epsilon(\mathbf{k}) = \mu = 0$), the quasi-particle energy vanishes now along the diagonals: $k_x = \pm k_y$ of the Brillouin zone.

Again at exact half-filling, another solution is possible, the so-called "mixed phase" of Ref. [120], corresponding to: $\phi = \pi/2$. Indeed, in such a case, the quasiparticle spectrum turns out to be given by:

$$E(\mathbf{k}) = J|\Delta|\sqrt{\cos^2(k_x a) + \cos^2(k_y a)} \tag{5.29}$$

and the condition (5.24) is satisfied in this case. The order parameter is now genuinely complex, corresponding to: $\Delta_y = i\Delta_x$ (another solution with: $\Delta_y = -i\Delta_x$ is also possible, but the two are related by the symmetry transformation (5.19)), and to:

$$\mathcal{U}(\mathbf{k}) = \Delta(\cos(k_x a) + i \cos(k_y a)) . \tag{5.30}$$

A fourth solution (not considered in Ref. [120]) is also possible at half-filling, one which we will call a "dimer" solution of the Baskaran-Zou-Anderson self-consistency equations, namely one corresponding to one of the two independent order parameters being zero. Indeed, if, say, we set: $\Delta_y = 0$ and: $\Delta_x = \Delta$, we obtain:

$$E(\mathbf{k}) = J|\Delta||\cos(k_x a)| . \tag{5.31}$$

Then, it is easy to check that the r.h.s. of Eq. (5.18b) consistently vanishes, while Δ is determined by:

$$1 = \frac{J}{M} \sum_{\mathbf{k}} \frac{\cos^2(k_x a)}{E(\mathbf{k})} \tanh\left(\frac{\beta E(\mathbf{k})}{2}\right) \tag{5.32}$$

or, more explicitly, by:

$$\Delta = \frac{1}{M} \sum_{\mathbf{k}} |\cos(k_x a)| \tanh\left[\frac{\beta J \Delta |\cos(k_x a)|}{2}\right] . \tag{5.33}$$

This "dimer" solution seems to describe a rather unphysical situation, namely one in which all the spins are coupled along one of the coordinate axes, while the coupling in the orthogonal direction vanishes. Although we have not checked this explicitly, we think that this situation may correspond to a local maximum of the free energy, that is, that it should be unstable against an arbitrarily small coupling in the orthogonal direction.

Kotliar [120] has discussed the transition temperatures associated to the "s-wave" and "d-wave" phases as a function of the dopant concentration δ. We refer to the original paper for details. Also, numerical estimates exhibited in Ref. [120] indicate that the "mixed" phase is (at half-filling) energetically favored with respect to the other ones at low temperatures.

Affleck and Marston [6, 47] have proposed a different mean field theory approach to the half-filled 2D Hubbard Hamiltonian which we will now review, mainly from our point of view. Let us start again (cf. Secs. 2 and 4) from the Hamiltonian:

$$\mathcal{H} = J \sum_{(ij)} \left[\mathbf{S}_i \cdot \mathbf{S}_j - \frac{\hat{n}_i \hat{n}_j}{4} \right]. \tag{5.34}$$

Then, using once more the identity (4.4), one can prove with some algebra that:

$$\mathbf{S}_i \cdot \mathbf{S}_j \equiv \frac{1}{2} \sum_{\alpha\beta} c_{i\alpha}^\dagger c_{i\beta} c_{j\beta}^\dagger c_{j\alpha} - \frac{\hat{n}_i \hat{n}_j}{4}. \tag{5.35}$$

At half-filling $\hat{n}_i \equiv 1 \ \forall i$, so, neglecting constants, we find:

$$\mathcal{H} = \frac{J}{2} \sum_{(ij)} \sum_{\alpha\beta} c_{i\alpha}^\dagger c_{i\beta} c_{j\beta}^\dagger c_{j\alpha}. \tag{5.36}$$

Affleck and Marston [6] have proposed that the spin indices in (5.36) be interpreted as "flavour" indices for a family of (independent) fermions, and then that the model be generalized to an arbitrary number $n \geq 2$ (n even) of fermions, renormalizing consequently the factor of ($J/2$) to (J/n) in the effective Hamiltonian (5.36). We will not follow here the analysis of the $n \to \infty$ limit performed in Ref. [6], but rather stick to the (physically interesting) $n = 2$ case, proceeding at an unrestricted (as far as the effect of the Gutzwiller projection goes) mean field level, which is an accurate description in the large n limit.

Partly following Ref. [6], we shall select the relevant order parameter as:

$$\chi_{ij} \equiv \chi_{ji}^* =: -\frac{J}{2} \sum_{\alpha} \langle c_{i\alpha}^\dagger c_{j\alpha} \rangle \tag{5.37}$$

for (i, j) n.n.'s, and zero otherwise.

A standard Hartree-Fock factorization yields then the effective Hamiltonian:

$$\mathcal{H}_{\text{eff}} = \sum_{(ij)} \left\{ \left[\sum_{\alpha} \chi_{ij} c_{j\alpha}^\dagger c_{i\alpha} + \text{h.c.} \right] + \frac{2}{J} |\chi_{ij}|^2 \right\}. \tag{5.38}$$

Remark: The choice (5.37) for the order parameter seems to be completely uncorrelated with the previous ones. That this is not actually completely so is a point that will be discussed in the next section.

Before proceeding, let us remark [6] that the Hamiltonian (5.36) is invariant under the local gauge transformation:

$$c_{j\sigma} \to \exp[i\theta_j] \cdot c_{j\sigma} \qquad (5.39)$$

with θ_j an arbitrary site-dependent function. This reflects the fact that:

$$[\![\mathcal{H}, \hat{n}_i]\!] = 0 \quad \forall i , \qquad (5.40)$$

that is, that the Hamiltonian (5.36) conserves the particle number at each lattice site. Due to the particular structure of the Hubbard Hamiltonian in the low energy sector, a quite familiar symmetry (the global U(1) symmetry corresponding to: $\theta_i = \text{const.} = \theta$ in (5.39)) of a many-body Hamiltonian conserving the total particle number is now promoted to the rôle of a local gauge symmetry.

Setting now:

$$\chi_{hk} = |\chi_{hk}| \cdot \exp(i\theta_{hk}) \qquad (5.41)$$

the phase of the order parameter transforms under the transformation (5.39) as:

$$\theta_{hk} \to \theta_{hk} + \theta_k - \theta_h . \qquad (5.42)$$

The transformation (5.42) is the discrete (lattice) version [29] of the transformation of a U(1) gauge field, and the phase of the order parameter can then be viewed as a U(1) lattice gauge field. The phase in itself is not a gauge-invariant quantity, but, e.g., the sum of the phases around an "elementary plaquette", namely:

$$\Delta\theta =: \theta_{hk} + \theta_{kl} + \theta_{lm} + \theta_{mh} \qquad (5.43)$$

(where h, k, l, and m are on the corners of a primitive cell of the square lattice) is a guage invariant quantity. $\Delta\theta$ can then be connected to observable effects.

The original square lattice can also be viewed as a centered cubic lattice (there is of course no distinction, in two space dimensions, between f.c.c. and b.c.c. lattices) with a unit cell of side: $a \cdot \sqrt{2}$ (see Fig. 9), having then the primitive lattice translation vectors along the diagonals of the original lattice. When viewed in this way, the lattice becomes "bipartite", i.e. it can be divided into two identical (simply cubic) interpenetrating sublattices, with the link variables χ_{ij} coupling only sites lying in different sublattices.

We can refer the new lattice to Cartesian coordinates X and Y which are rotated by $\pi/4$ with respect to the original coordinates x and y. As, in this process, the primitive cell of the lattice has been doubled, the corresponding Brillouin zone will be halved, and will correspond to the hatched area in Fig. 10.

The order parameter (5.37) can take in principle four independent values, corresponding to the four bonds emanating from any given lattice site. The resulting pattern of bonds in the bipartite lattice is again shown in Fig. 9, where the four bonds, labeled χ_1 to χ_4, are depicted with an arrow, which is intended to be a reminder of the fact that as, generically: $\chi_{ij} \neq \chi_{ji}$, the bonds do have now a directional character. Note that, with the pattern of bonds depicted in Fig. 9, the system has again full translational symmetry in the bipartite lattice.

Let us indicate with $c_{i\sigma}^\dagger$ and $d_{i\sigma}^\dagger$ the creation operators for electrons in the two sublattices of Fig. 9. We are then considering formally the electrons in the two sublattices as two independent species of fermions. With reference to the geometry depicted in Fig. 11, and to the pattern of bonds of Fig. 9, the effective Hamiltonian (5.38) can be written as:

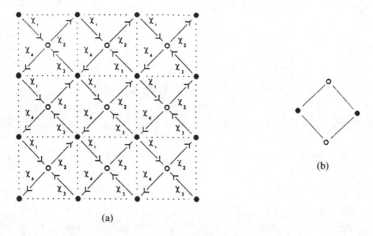

Fig. 9. a) The array of n.n. bonds corresponding to the Affleck-Marston mean-field solution. b) The original 2D square lattice unit cel.

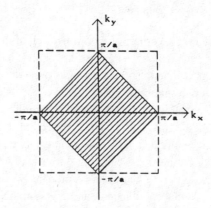

Fig. 10. The reduced Brillouin zone (hatched area) for the doubling of the lattice required in the approach of Affleck and Marston.

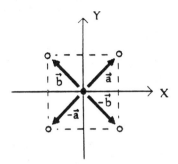

Fig. 11. Lattice vectors for the geometry discussed in Sec. 5.

$$\mathcal{H}_{\text{eff}} = \sum_{j,\alpha} \{\chi_1 c_{j\alpha}^\dagger d_{j-\mathbf{b},\alpha} + \chi_2^* c_{j\alpha}^\dagger d_{j-\mathbf{a},\alpha} + \chi_3 c_{j\alpha}^\dagger d_{j+\mathbf{b},\alpha} + \chi_4^* c_{j\alpha}^\dagger d_{j+\mathbf{a},\alpha} + \text{h.c.}\} \tag{5.44}$$

where, for the time being, we have omitted the last term in (5.38), and:

$$\chi_1 =: -\frac{J}{2} \sum_\alpha \langle d_{j-\mathbf{b},\alpha}^\dagger c_{j\alpha} \rangle \tag{5.45a}$$

$$\chi_2 =: -\frac{J}{2} \sum_\alpha \langle c_{j\alpha}^\dagger d_{j-\mathbf{a},\alpha} \rangle \tag{5.45b}$$

$$\chi_3 =: -\frac{J}{2} \sum_\alpha \langle d_{j+\mathbf{b},\alpha}^\dagger c_{j\alpha} \rangle \tag{5.45c}$$

$$\chi_4 =: -\frac{J}{2} \sum_\alpha \langle c_{j\alpha}^\dagger d_{j+\mathbf{a},\alpha} \rangle . \tag{5.45d}$$

The sums in (5.44) are to be understood as sums over the sites of one of the two sublattices of the bipartite lattice.

Introducing, in the usual way, the Fourier transforms of the Fermion operators, (5.44) becomes:

$$\mathcal{H}_{\text{eff}} = \sum_{\mathbf{k},\alpha} [\lambda(\mathbf{k}) c_{\mathbf{k}\alpha}^\dagger d_{\mathbf{k}\alpha} + \text{h.c.}] \tag{5.46}$$

where:

$$\lambda(\mathbf{k}) =: \chi_1 e^{ik_x a} + \chi_2^* e^{-ik_y a} + \chi_3 e^{-ik_x a} + \chi_4^* e^{ik_y a} \tag{5.47}$$

and the **k**-summation is on the reduced Brillouin zone.

Introducing new fermion operators $\alpha_{\mathbf{k}\sigma}$ and $\beta_{\mathbf{k}\sigma}$ related to the c's and d's by the canonical transformation:

$$c_{\mathbf{k}\sigma} = \frac{1}{\sqrt{2}} \left[\alpha_{\mathbf{k}\sigma} - \frac{\lambda(\mathbf{k})}{|\lambda(\mathbf{k})|} \beta_{\mathbf{k}\sigma} \right]$$

$$d_{\mathbf{k}\sigma} = \frac{1}{\sqrt{2}} \left[\frac{\lambda(\mathbf{k})^*}{|\lambda(\mathbf{k})|} \alpha_{\mathbf{k}\sigma} + \beta_{\mathbf{k}\sigma} \right] \quad (5.48)$$

the Hamiltonian (5.46) acquires the diagonal form:

$$\mathcal{H}_{\text{eff}} = \sum_{\mathbf{k},\alpha} E(\mathbf{k})[\alpha_{\mathbf{k}\alpha}^\dagger \alpha_{\mathbf{k}\alpha} - \beta_{\mathbf{k}\alpha}^\dagger \beta_{\mathbf{k}\alpha}] \quad (5.49)$$

with quasi-particle energies $E(\mathbf{k})$ given by:

$$E(\mathbf{k}) =: |\lambda(\mathbf{k})| . \quad (5.50)$$

Working out explicitly the expectation values on the r.h.s. of Eqs. (5.45), we obtain the following set of coupled nonlinear self-consistency equations:

$$\chi_1 = \frac{J}{M} \sum_{\mathbf{k}} \exp[-ik_x a] \frac{\lambda(\mathbf{k})}{2|\lambda(\mathbf{k})|} \tanh\left(\frac{\beta|\lambda(\mathbf{k})|}{2}\right) \quad (5.51a)$$

$$\chi_2 = \frac{J}{M} \sum_{\mathbf{k}} \exp[-ik_y a] \frac{\lambda(\mathbf{k})^*}{2|\lambda(\mathbf{k})|} \tanh\left(\frac{\beta|\lambda(\mathbf{k})|}{2}\right) \quad (5.51b)$$

$$\chi_3 = \frac{J}{M} \sum_{\mathbf{k}} \exp[ik_x a] \frac{\lambda(\mathbf{k})}{2|\lambda(\mathbf{k})|} \tanh\left(\frac{\beta|\lambda(\mathbf{k})|}{2}\right) \quad (5.51c)$$

$$\chi_4 = \frac{J}{M} \sum_{\mathbf{k}} \exp[ik_y a] \frac{\lambda(\mathbf{k})^*}{2|\lambda(\mathbf{k})|} \tanh\left(\frac{\beta|\lambda(\mathbf{k})|}{2}\right) \quad (5.51d)$$

where $\lambda(\mathbf{k})$ is given by Eq. (5.47).

A Digression on: Gauge Invariance and its Consequences

We have already discussed the U(1) gauge invariance of the Hamiltonian (5.36) (see Eq. (5.39)). We shall explore here more closely some of its consequences.

In a fixed gauge, the fermion operators in any one of the two sublattices are connected to their Fourier transformed counterparts by:

$$c_j = \sqrt{\frac{2}{M}} \sum_{\mathbf{q}} e^{i\mathbf{q}\cdot\mathbf{R}_j} c_{\mathbf{q}} ; \quad c_{\mathbf{q}} = \sqrt{\frac{2}{M}} \sum_{j} e^{-i\mathbf{q}\cdot\mathbf{R}_j} c_j \quad (5.52)$$

and similarly for the d operators. Spin indices will be dropped for the time being, as they play no role in the discussion. The wave vector $\mathbf{q}$ in (5.52) is assumed to run on the (reduced) Brillouin zone, which can be identified with the hatch area of Fig. 10, while the j sum is over the Bravais lattice of the bipartite lattice of Fig. 9.

Defining now:

$$\tilde{c}_j =: \exp[i\theta_j] \cdot c_j \qquad (5.53)$$

we will define again the Fourier transformed operators as:

$$\tilde{c}_\mathbf{k} =: \sqrt{\frac{2}{M}} \sum_j e^{-i\mathbf{k} \cdot \mathbf{R}_j} \tilde{c}_j \qquad (5.54)$$

where the wave vector $\mathbf{k}$ runs again over the Brillouin zone. Expressing the $\tilde{c}_j$'s via Eq. (5.53), and using the first of Eqs. (5.52), we find:

$$\tilde{c}_\mathbf{k} = \sum_\mathbf{q} \phi_c(\mathbf{q} - \mathbf{k}) \cdot c_\mathbf{q} \qquad (5.55)$$

where:

$$\phi_c(\mathbf{q} - \mathbf{k}) = \frac{2}{M} \sum_j \exp\{i[(\mathbf{q} - \mathbf{k}) \cdot \mathbf{R}_j + \theta_j]\} \ . \qquad (5.56)$$

A similar kernel (with *a priori* different phases) ϕ_d will be defined for the d lattice. One can easily check that the kernel ϕ ($\phi = \phi_c$ or $\phi = \phi_d$) satisfies:

$$\sum_\mathbf{k} \phi^*(\mathbf{q} - \mathbf{k})\phi(\mathbf{q}' - \mathbf{k}) = \delta_{\mathbf{q},\mathbf{q}'} \ ; \quad \sum_\mathbf{q} \phi^*(\mathbf{q} - \mathbf{k})\phi(\mathbf{q} - \mathbf{k}') = \delta_{\mathbf{k},\mathbf{k}'} \quad (5.57)$$

which ensures the unitarity of the transformation defined by Eq. (5.55). The Fourier transformed operators are therefore unitarily related, as expected ((5.53) is of course a unitary transformation).

If we reduce further the gauge group to the (more familiar) global U(1) group ($\theta_j \equiv$ const. $\equiv \theta$), then: $\phi(\mathbf{q} - \mathbf{k}) = \exp(i\theta) \cdot \delta_{\mathbf{k},\mathbf{q}}$, and the transformation (5.55) becomes essentially trivial. Actually, two independent U(1)'s are allowed, one for each sublattice, and we end up with a (global) U(1) × U(1) symmetry.

Let us consider now what are the effects of the (local) U(1) symmetry on the solutions we have just found (Eqs. (5.38) to (5.51)).

Under the action of the gauge group, the four bonds emanating from a given site j change as follows (cf. also Eqs. (5.45)):

$$\chi_1 \to \chi_1 \exp[i(\theta_j - \theta_{j-\mathbf{b}})] \qquad (5.58a)$$

$$\chi_2 \to \chi_2 \exp[i(\theta_{j-\mathbf{a}} - \theta_j)] \qquad (5.58b)$$

$$\chi_3 \to \chi_3 \exp[i(\theta_j - \theta_{j+\mathbf{b}})] \qquad (5.58c)$$

$$\chi_4 \to \chi_4 \exp[i(\theta_{j+\mathbf{a}} - \theta_j)] \ . \qquad (5.58d)$$

Alternatively, the four bonds along the perimeter of an elementary "plaquette" with one corner at site j (Fig. 12) change as:

$$\chi_1 \to \chi_1 \exp[i(\theta_j - \theta_{j-\mathbf{b}})] \qquad (5.59\text{a})$$

$$\chi_2 \to \chi_2 \exp[i(\theta_{j-\mathbf{b}} - \theta_{j+\mathbf{a}-\mathbf{b}})] \qquad (5.59\text{b})$$

$$\chi_3 \to \chi_3 \exp[i(\theta_{j+\mathbf{a}-\mathbf{b}} - \theta_{j+\mathbf{a}})] \qquad (5.59\text{c})$$

$$\chi_4 \to \chi_4 \exp[i(\theta_{j+\mathbf{a}} - \theta_j)] . \qquad (5.59\text{d})$$

If we want to keep the pattern of bonds of Fig. 9 invariant under displacements in the bipartite lattice, i.e.:

$$\chi_{ij} = \chi_{i+\xi, j+\xi}, \qquad \forall \xi \qquad (5.60)$$

where (cf. Fig. 11):

$$\xi =: n(\mathbf{a} + \mathbf{b}) + m(\mathbf{a} - \mathbf{b}) \qquad (5.61)$$

with n, m integers is any vector in the Bravais lattice of the bipartite lattice of Fig. 9, then the U(1) gauge symmetry should be broken down to the subgroup generated by phases θ_j satisfying:

$$\theta_j - \theta_{j\pm\mathbf{x}} = \theta_{j+\xi} - \theta_{j\pm\mathbf{x}+\xi}; \qquad \mathbf{x} = \mathbf{a} \quad \text{or} \quad \mathbf{x} = \mathbf{b}. \qquad (5.62)$$

It can be checked that, with this requirement, Eqs. (5.58b,c) become identical with (5.59b,c) respectively. It is also not difficult to convince oneself that the only way to enforce the constraint of Eq. (5.62) is to select at will one given site j and to fix the four phases at the corners of an elementary plaquette, i.e. at sites (see Fig. 12) $j + \mathbf{a}$, $j - \mathbf{b}$, $j + \mathbf{a} - \mathbf{b}$, and then to generate all the other phases by translations with lattice vectors of the form (5.61). But then we obtain a reduction of the gauge group to U(1) × U(1) × U(1) × U(1), or, quotienting with the previous U(1) × U(1), a residual U(1) × U(1) gauge group.

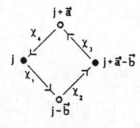

Fig. 12. Array of bonds along an elementary plaquette (Affleck and Marston's approach).

In this case, setting:

$$\phi =: \theta_j - \theta_{j-\mathbf{b}} \qquad (5.63)$$

we easily find:

$$\chi_i \to \chi_i e^{i\phi}, \quad i = 1, 3; \qquad \chi_i \to \chi_i e^{-i\phi}, \quad i = 2, 4; \qquad (5.64)$$

whence:

$$\lambda(\mathbf{k}) \to \lambda(\mathbf{k}) e^{i\phi} \qquad (5.65)$$

and the quasiparticle energies (5.50) are left unaltered as well.

It appears therefore as if we were facing here a spontaneous breaking of the U(1) gauge symmetry down to a global U(1) × U(1) × U(1) × U(1) (or U(1) × U(1)) one. However, such breakings are forbidden by Elitzur's theorem [64]. The correct meaning of what we have been doing up to now is that it is possible to fix the gauge in such a way that the solutions of the self consistency equations satisfy Eq. (5.60) (note that this is by no means an *a priori* obvious nor a trivial statement). However, as the physics must be independent from the gauge choice, all the solutions obtained by "gauging" (independently on each plaquette) those of Eqs. (5.51) must be considered on equal footing. Only gauge invariant quantities can have physical meaning, and one has to be careful to attribute physical significance only to quantities of the latter kind. In particular, as will be seen also shortly below in specific examples, the quasiparticle energies (5.50) are in general gauge dependent quantities.

We discuss now various solutions of the self-consistency equations (5.51).

A solution is easily found, in this case as well as in the previous one, in which all the four χ's are equal and real:

$$\chi_i = \chi; \quad i = 1, \ldots 4; \quad \chi > 0. \qquad (5.66)$$

Indeed, imposing (5.66) and exploiting the ensuing symmetries of the problem, one easily finds a single equation for χ, namely:

$$\chi = \frac{J}{2M} \sum_{\mathbf{k}} |\gamma(\mathbf{k})| \cdot \tanh(\beta \chi |\gamma(\mathbf{k})|). \qquad (5.67)$$

This solution closely resembles the "s-wave" solution of Kotliar at half-filling. However, the authors of Ref. [6] did not find a solution of the form (5.66) in their numerical search of the stable minima of the free energy, so we conclude that (5.66) must actually correspond to a (local) maximum of the free energy. Equation (5.56) corresponds of course to $\Delta\theta = 0$ from Eq. (5.43), that is to a zero total phase variation for a circuit around an elementary plaquette. All the solutions with $\Delta\theta = 0$ fall then into the same gauge class of (5.66).

Another solution, called the "dimer" solution in Ref. [6], corresponds to only one of the χ_i's, say χ_1, being nonzero. Note that, if this is the case (or whenever at least one of the χ_i's vanishes), the total phase variation (5.43) becomes meaningless. Setting then:

$$\chi_2 = \chi_3 = \chi_4 = 0 ; \qquad \chi_1 = \chi > 0 \tag{5.68}$$

(a positive χ_1 can always be obtained with the aid of a gauge transformation) one finds, again with some simple algebra:

$$\chi = \frac{J}{2M} \sum_\mathbf{k} \tanh\left(\frac{\beta\chi}{2}\right) \equiv \frac{J}{4} \tanh\left(\frac{\beta\chi}{2}\right). \tag{5.69}$$

Hence $\chi = J/4$ at $T = 0$, and the transition temperature is (setting Boltzmann's constant to unity):

$$T_c = \frac{J}{8}. \tag{5.70}$$

The dimer phase "melts" at $T = T_c$. This dimer phase (also called the "Peierls" phase by Affleck and Marston) has been found numerically in Ref. [6] to correspond to a stable local minimum of the free energy. The geometry of the dimer phase is also discussed in Ref. [6].

We discuss next solutions corresponding to all the four χ_i's having equal magnitudes, and possibly different phases. Let us remark, first of all, that Eqs. (5.51) support, for any given solution, also the solutions obtained by the following interchanges:

$$\chi_1 \Leftrightarrow \chi_2^* ; \qquad \chi_3 \Leftrightarrow \chi_4^* \tag{5.71}$$

$$\chi_1 \Leftrightarrow \chi_3 ; \qquad \text{and/or} \qquad \chi_2 \Leftrightarrow \chi_4 \tag{5.72}$$

(5.71) can be obtained by interchanging k_x with k_y in the sums, while (5.72) is obtained by sending: $k_x \to -k_x$ and/or: $k_y \to -k_y$.

Solutions with $\chi_1 = \chi_2^* = \chi_3 = \chi_4^*$ correspond to $\Delta\theta = 0$, and, for the reasons already discussed, will not be considered here.

We next consider solutions of the form:

$$\chi_1 = \chi_3 = \chi > 0 ; \qquad \chi_2 = \chi_4 = \chi e^{i\phi} . \tag{5.73}$$

The discussion of this class of solutions follows closely that of Kotliar's "mixed phase". Equation (5.73) leads to:

$$\lambda(\mathbf{k}) = 2\chi \cdot (\cos(k_x a) + e^{-i\phi} \cos(k_y a)) . \tag{5.74}$$

This implies that $E(\mathbf{k}) = |\lambda(\mathbf{k})|$ depends only on $\cos(\phi)$, as in Eq. (5.21)). After manipulations quite analogous to those leading to Eqs. (5.22)–(5.23), we end up with the following complex equation for χ and ϕ:

$$\chi = \frac{J}{2M} \sum_{\mathbf{k}} \cos(k_x a) \frac{\lambda(\mathbf{k})}{|\lambda(\mathbf{k})|} \tanh\left(\frac{\beta|\lambda(\mathbf{k})|}{2}\right). \tag{5.75}$$

We obtain then, just as previously:

$$\sin\phi \sum_{\mathbf{k}} \frac{\cos(k_x a)\cos(k_y a)}{|\lambda(\mathbf{k})|} \tanh\left(\frac{\beta|\lambda(\mathbf{k})|}{2}\right) = 0. \tag{5.76}$$

This equation is satisfied for $\phi = 0$, π and $\phi = \pi/2$. The first two solutions are gauge equivalent to (5.66) (they correspond to $\Delta\theta = 0, 2\pi$). The third has

$$\Delta\theta = \pi \tag{5.77}$$

and quasiparticle energies

$$E(\mathbf{k}) = 2\chi\sqrt{\cos^2(k_x a) + \cos^2(k_y a)}. \tag{5.78}$$

The same result has been obtained in Ref. [6] starting from the gauge equivalent (but more symmetric) ansatz:

$$\chi_i = \chi \cdot \exp\left[\frac{i\pi}{4}\right]; \quad i = 1, \ldots, 4. \tag{5.79}$$

It has been called the "flux phase" by Affleck and Marston.

The quasiparticle energies have the same form as in Kotliar's "mixed phase". They vanish at $\mathbf{k} = (\pm\pi/2a, \pm\pi/2a)$. However, only two of these points (say $(\pi/2a, \pm\pi/2a)$) exist as independent points in the reduced Brillouin zone. The other two are obtained from these by translations by one reciprocal lattice vector.

If the total particle number has to be conserved, then only particle-hole excitations can be permitted, and we see that we have two gapless mode, at $\mathbf{k} = (0, 0)$ and at $\mathbf{k} = (0, \pi/a)$ respectively.

A gauge equivalent way to derive the flux phase is to "concentrate" all the phases on a single bond, that is to make the ansatz:

$$\chi_1 = \chi_2 = \chi_3 = \chi > 0; \quad \chi_4 = \chi \cdot e^{i\phi}. \tag{5.80}$$

The discussion can be pursued for general values of ϕ but, as the final conclusion will turn out to be the same, that is, that the only nontrivial allowed value is $\phi = \pi$, we will set directly $\phi = \pi$ in what follows. It turns out that the solution exists, but now the quasiparticle energies are given by

$$E(\mathbf{k}) = 2\chi\sqrt{\cos^2(k_x a) + \sin^2(k_y a)}. \tag{5.81}$$

Now, however, the zeros of $E(\mathbf{k})$ are at $k = (\pm\pi/2, 0)$ (plus equivalent points obtained by reciprocal lattice translations). Comparison of (5.78) and (5.81) exhibits clearly the gauge dependent nature of the quasiparticle spectrum. However, note that, considering particle-hole excitations, we obtain again gapless modes at $\mathbf{k} = (0, 0)$ and at $\mathbf{k} = (\pi/a, 0)$, which is equivalent to $(0, \pi/a)$. The location of the gapless particle-hole excitations seems therefore to be a gauge independent notion.

Energetically, the dimer phase seems to be favored with respect to the flux phase at exactly half-filling, according to the numerical analysis of Ref. [6]. In the same paper, Affleck and Marston have also extended the analysis below half-filling, adding a hopping term to the Hamiltonian, which becomes then a generalized tight-binding Hamiltonian will an effective hopping term given by $\chi_{ij} + t_{ij} (t_{ij} \simeq t)$. They find a fourth stable phase (called the "kite" phase) at intermediate dopings, characterized by the χ's being real, and: $\chi_1 = \chi_2 \neq \chi_3 = \chi_4$. We will not however discuss it here. We would like only to comment that, due to the various terms that have been neglected in deriving the effective Hamiltonian, and whose neglect is justified only at half-filling, straightforward extensions of the formalism below half-filling do not seem to be on such firm bases as the results discussed up to now.

We close this rather lengthy section with a brief account of the path-integral approach to flux phases [67].

A number of authors [6, 8, 38–40], have expressd the partition function of the Hubbard Hamiltonian at and near half filling as a path integral over some appropriate Hubbard-Stratonovich field (see Sec. 1 for a discussion of the use of the Hubbard-Stratonovich identity to set up path integrals). The main purpose of this activity seems to have been to show that, after the fermionic degrees of freedom have been integrated out, the effective free energy in the path integral (see below for details) has an expansion in terms of Wegner-Wilson-type loops [118]. Then, and to some extent, this allows to set up an analogy between the strong-coupling limit of the Hubbard model and lattice QCD.

The path-integral formulation can be derived following the method outlined in Sec. 1. We will fix our attention on flux phases, and hence on the form of the Hamiltonian discussed by Affleck and Marston in Ref. [6]. Also, for the time being, we will concentrate our attention on the half-filling case.

Neglecting terms which are constant at half-filling, the Hamiltonian (5.36) can be rewritten as:

$$\mathcal{H} = -\frac{J}{2} \sum_{(ij)} \sum_{\alpha\beta} c_{i\alpha}^\dagger c_{j\alpha} c_{j\beta}^\dagger c_{i\beta} . \tag{5.82}$$

Defining:

$$\hat{\chi}_{ij} =: \sum_\alpha c_{i\alpha}^\dagger c_{j\alpha} ; \qquad \hat{\chi}_{ij}^\dagger = \hat{\chi}_{ji} \tag{5.83}$$

(note that $\langle \hat{\chi}_{ij} \rangle$ differs by a numerical factor from the analogous quantity defined in Eq. (5.37)). (5.82) can be rewritten as:

$$\mathcal{H} = -\frac{J}{2} \sum_{(ij)} \hat{\chi}_{ij} \hat{\chi}_{ij}^{\dagger} \tag{5.84}$$

and it is to this Hamiltonian that the Hubbard-Stratonovich identity has to be applied.

We will limit ourselves here to a discussion of the static approximation to the path-integral, i.e., we will restrict the integration over paths to Hubbard-Stratonovich fields which do not depend on the (imaginary) time. The partition function at half-filling and in the static approximation is then given by:

$$\mathcal{Z} = \int \prod_{(ij)} d^2 \mathcal{U}_{ij} \exp\left[-\pi \sum_{(ij)} |\mathcal{U}_{ij}|^2\right] \operatorname{Tr}\{\exp[-\beta \mathcal{H}_{\text{eff}}]\} \tag{5.85}$$

where the trace has to be taken in the subspace of half-filled states (which has dimension 2^N), $\mathcal{U}_{ij}$ is the (complex) Hubbard-Stratonovich field coupled to $\chi_{ij}^{\dagger}$ obeying (cf. the second of (5.83)) $\mathcal{U}_{ji} = \mathcal{U}_{ij}{}^*$, and the ($\mathcal{U}$-dependent) effective one-body Hamiltonian $\mathcal{H}_{\text{eff}}$ is given by:

$$\mathcal{H}_{\text{eff}} = \sqrt{\frac{\pi J}{2\beta}} \sum_{(ij)} \{\mathcal{U}_{ij}^* \hat{\chi}_{ij} + \text{h.c.}\} . \tag{5.86}$$

In view of the symmetry relations obeyed by both χ_{ij} and $\mathcal{U}_{ij}$, we can also write:

$$\mathcal{H}_{\text{eff}} = \sqrt{\frac{\pi J}{2\beta}} \sum_{\langle ij \rangle} \mathcal{U}_{ij}^* \hat{\chi}_{ij} \equiv \sqrt{\frac{\pi J}{2\beta}} \sum_{\langle ij \rangle} \mathcal{U}_{ji} \hat{\chi}_{ij} . \tag{5.87}$$

Note that the sums are now over *ordered* pairs.

It is convenient to introduce a matrix $\tilde{\mathcal{U}}_{ij}$ defined as:

$$\tilde{\mathcal{U}}_{ij} = \mathcal{U}_{ij} \text{ for } (i,j) \text{ n.n.}, \quad \text{and} \quad \tilde{\mathcal{U}}_{ij} = 0 \text{ otherwise} . \tag{5.88}$$

In this way we can get formally rid of the summation restrictions, and write:

$$\mathcal{H}_{\text{eff}} = \tilde{J} \sum_{ij} \tilde{\mathcal{U}}_{ij} \chi_{ji} \equiv \tilde{J} \operatorname{Sp}\{\tilde{\mathcal{U}} \chi\} ; \quad \tilde{J} =: \sqrt{\frac{\pi J}{2\beta}} \tag{5.89}$$

where $\operatorname{Sp}\{.\}$ stand for a trace of whatever is inside the curly brackets, considered as a (generalized) matrix in the site indices. As $\tilde{\mathcal{U}} = \|\tilde{\mathcal{U}}_{ij}\|$ is a hermitean matrix, $\mathcal{H}_{\text{eff}}$ will be hermitean as well.

The quantum-mechanical trace can be evaluated explicitly for Hamiltonians bilinear in fermion operators like (5.89) [66, 135]. Defining the thermodynamic Green functions as in Eqs. (5.4)–(5.6), we define further:

$$\mathbb{G}_{ij}(\tau) =: \sum_{\sigma} \mathcal{G}_{\sigma}^{ij}(\tau) \tag{5.90}$$

and denote by $\mathcal{G}_{ij}(\omega_n)$ its Fourier transform (cf. Eqs. (5.6, 7)). One can then easily prove [67] that the matrix $\mathcal{G}_{ij}(\omega_n) \equiv \|\mathcal{G}_{ij}(\omega_n)\|$ is given by:

$$\mathcal{G}(\omega_n) = \mathcal{G}^{(0)}(\omega_n)[\mathbb{1} - \mathcal{K} \cdot \mathcal{G}^{(0)}(\omega_n)]^{-1} \equiv [\mathcal{G}^{(0)-1}(\omega_n) - \mathcal{K}]^{-1} \quad (5.91)$$

where:

$$\mathcal{G}^{(0)}{}_{ij}(\omega_n) = \frac{2\delta_{ij}}{i\omega_n}; \qquad \mathcal{K}_{ij} = \frac{\tilde{J}}{2\hbar} \tilde{\mathcal{U}}_{ij} \quad (5.92)$$

The spectrum of $\mathcal{H}_{\text{eff}}$ is given by the poles of the analytic continuation of $\mathcal{G}$ from $i\hbar\omega_n$ to a complex variable ζ. It will coincide with the spectrum of $\hbar\mathcal{K}$, and, as $\mathcal{K}$ is hermitean, it is clear that the spectrum will be entirely real, as it should be.

Let us return now to (5.85). Writing the path integral as:

$$\mathfrak{L} = \int \prod_{(ij)} d^2 \mathcal{U}_{ij} \exp\left[-\pi \sum_{(ij)} |\mathcal{U}_{ij}|^2\right] \mathfrak{L}(\{\mathcal{U}\}) \quad (5.93)$$

where:

$$\mathfrak{L}(\{\mathcal{U}\}) =: \mathrm{Tr} \exp[-\beta \mathcal{H}_{\text{eff}}] \quad (5.94)$$

one can prove [67] that:

$$\mathfrak{L}(\{\mathcal{U}\}) = 2^N \exp\left[-2 \sum_n e^{i\omega_n 0^+} \mathrm{Sp} \ln(\mathbb{1} - \mathcal{K}\mathcal{G}^{(0)}(\omega_n))\right]. \quad (5.95)$$

The factor 2^N contributes an additive term of $N \ln 2$ to the logarithm of the partition function. Such a term is nothing but the zero-temperature entropy associated with the degeneracy of the ground state. As it does not affect any physical quantity, it will be dropped henceforth.

Expanding formally the argument of the exponential in (5.95), we obtain:

$$\sum_n e^{i\omega_n 0^+} \mathrm{Sp} \ln[\mathbb{1} - \mathcal{K}\mathcal{G}^{(0)}(\omega_n)] = \sum_{p=1}^{\infty} \frac{\mathrm{Sp}(\mathcal{K}^p)}{p} \sum_n e^{i\omega_n 0^+} [\mathcal{G}^{(0)}(\omega_n)]^p \quad (5.96)$$

where the factor δ_{ij} has been removed from $\mathcal{G}^{(0)}$, which is now a scalar. As $\mathcal{G}^{(0)} \propto (i\omega_n)^{-1}$, the convergence factor $\exp[i\omega_n 0^+]$ is actually needed only in the first term of the expansion, and can be dropped in higher order terms. But, as $\mathcal{K}$ has no diagonal terms, $\mathrm{Sp}\,\mathcal{K} = 0$, and this takes care of the $p = 1$ term of the expansion. Moreover, one can easily convince oneself that

$$\sum_n [\mathcal{G}^{(0)}(\omega_n)]^{2k+1} \equiv 0 \quad \forall k \neq 0 \quad (5.97)$$

and so only traces of *even* powers of $\mathcal{K}$ will survive in the sum.

Remarks: 1) It is tempting to visualize $Sp(\mathcal{H}^p)$ as a sum over all the possible oriented loops in the lattice composed of p distinct n.n. bonds (and p distinct vertices, of course), with a factor of $\mathcal{H}_{ij}$ associated with the bond going from site i to site j. To every such loop there corresponds in the sum the oppositely oriented loop in which the orientation of the bonds is reversed and $\mathcal{H}_{ij}$ is replaced by $\mathcal{H}_{ji} = \mathcal{H}^*_{ij}$. Therefore, the sum of the contributions from such pairs of loops is real, and can be considered as a Wilson-Wegner loop of order p.

The situation is however not so simple. Taking, for example, the terms with $p = 2$, we see that

$$Sp(\mathcal{H}^2) = \left(\frac{\tilde{J}}{\hbar}\right)^2 \sum_{(ij)} |\tilde{\mathcal{U}}_{ij}|^2 \ . \tag{5.98}$$

The $p = 2$ term is therefore a "degenerate" loop, corresponding to a single bond being traced back and forth. The same thing will happen in higher order terms, whereby we can "decorate" a loop by attaching to each one of its vertices one bond or a string of bonds each one of which is traversed back and forth an arbitrary number of times and in any order. For example, if we expand up to fourth order, we obtain, besides (5.98) and neglecting overall multiplicative constants:

$$Sp(\tilde{\mathcal{U}}^4) = \sum_{(ij)} |\mathcal{U}_{ij}|^4 + \sum_{(ijk)} |\mathcal{U}_{ij}|^2 |\mathcal{U}_{jk}|^2 + 2 \operatorname{Re} \sum_{(ijkl)} \mathcal{U}_{ij} \mathcal{U}_{jk} \mathcal{U}_{kl} \mathcal{U}_{li} \tag{5.99}$$

where (ijk) means that i is n.n. to j, j to k and they are all different from each other, while $(ijkl)$ means that i is n.n. to j, j to k, k to l and l to i, the four indices being again all different from each other. Therefore, the first two terms in (5.99) are seen to arise from "decorating" the second order contribution with a bond traversed twice, and attaching the bond to (5.98) in the two possible ways. Only the last term corresponds to a genuine fourth order loop (i.e., in this case, to the contribution from an elementary plaquette in the lattice).

If we want to perform a partial resummation of the series (5.96) (restricted to even p's) in order to obtain a genuine expansion in terms of loops (which seems to be the spirit of the Wegner-Wilson-type of expansions that have been advocated in the literature), we have to renormalize each loop by attaching to each of its vertices arbitrary strings of single bonds, each bond carrying a factor of the form $|\mathcal{U}_{ij}|^2$. This looks like a formidable program to carry on systematically to all orders, and one we will not pursue here. It seems reasonable to assume that the effect of the renormalization, which does not affect the phases of the $\mathcal{U}_{ij}$'s, should be that of renormalizing their amplitudes away from their nominal values.

ii) Under the $\mathbb{U}(1)$ gauge transformation:

$$c_{j\sigma} \to \exp[i\theta_j] c_{j\sigma} \tag{5.100}$$

$\mathcal{H}_{ij}$ transforms as:

$$\mathcal{H}_{ij} \to \mathcal{H}_{ij} \exp[i(\theta_j - \theta_i)] \ . \tag{5.101}$$

The product of the $\mathcal{K}_{ij}$'s along a closed loop is then manifestly a gauge invariant quantity, and so is, *a fortiori*, $\mathrm{Sp}(\mathcal{K}^p)$.

iii) Equation (5.97) is no more valid away from half filling, as then $\mathcal{G}^{(0)}(\omega_n)$ does not have the simple form (5.92) anymore. Whether or not "odd" loops will appear in the expansion will depend then both on the topology of the lattice and the structure of $\mathcal{K}$. If $\mathcal{K}_{ij}$ is taken to be nonzero only for i, j nearest neighbors, as we are doing here, then "odd" loops will be allowed for a triangular lattice but not for a square lattice. For the latter only "even" loops will be allowed, no matter whether (5.97) holds or not. However, if one allows nonzero $\mathcal{K}_{ij}$'s also for next-nearest neighbors, as done, e.g., in Ref. [179], then "odd" loops will be allowed, away from half-filling, for the square lattice as well. We stress however that Eq. (5.97) will take care in any case of canceling the latter at exact half filling.

We now turn to a more detailed evaluation of (5.96). For even p, $p = 2k$, we have:

$$\sum_n [\mathcal{G}^{(0)}(\omega_n)]^{2k} = (-)^k 2^{2k} \left(\frac{\hbar\beta}{\pi}\right)^{2k} \sum_n (2n+1)^{-2k} . \tag{5.102}$$

The sum on the r.h.s. of (5.102) can be evaluated in terms of the Riemann zeta function. Indeed [1]:

$$\lambda_k =: \sum_{n=-\infty}^{\infty} \frac{1}{(2n+1)^{2k}} = 2(1 - 2^{-2k})\zeta(2k) \tag{5.103}$$

where ζ is the Riemann zeta function. Explicitly, and with accuracy up to the third decimal point [1]: $\lambda_1 = 1.123$, $\lambda_2 = 1.013$, $\lambda_3 = 1.001$, and, to the same accuracy, $\lambda_k = 1.000$ for $k > 3$.

Collecting results, we find:

$$2 \sum_n e^{i\omega_n 0^+} \mathrm{Sp} \ln[\![\mathbb{1} - \mathcal{K}\mathcal{G}^{(0)}(\omega_n)]\!] = \sum_{k=1}^{\infty} \frac{(-)^k \lambda_k}{k} \left(\frac{c}{\pi}\right)^{2k} \mathrm{Sp}(\tilde{\mathcal{U}}^{2k}) \tag{5.104}$$

where: $c = \sqrt{\pi\beta J/2}$.

Equation (5.93) can now be rewritten as:

$$\mathcal{L} = \int \prod_{(ij)} d^2 \mathcal{U}_{ij} \exp[\![-\beta \mathcal{F}[\tilde{\mathcal{U}}]]\!] \tag{5.105}$$

where the effective free energy $\mathcal{F}[\tilde{\mathcal{U}}]$ is given by:

$$\mathcal{F}[\tilde{\mathcal{U}}] = \frac{\pi}{\beta} \sum_{(ij)} |\mathcal{U}_{ij}|^2 + \frac{1}{\beta} \sum_{k=1}^{\infty} \frac{(-)^k \lambda_k}{k} \mathrm{Sp}\left(\frac{c\tilde{\mathcal{U}}}{\pi}\right)^{2k}$$

$$\equiv \frac{1}{\beta} \mathrm{Sp}\left\{\pi\tilde{\mathcal{U}}^2 + \sum_{k=1}^{\infty} \frac{(-)^k \lambda_k}{k} \left(\frac{c\tilde{\mathcal{U}}}{\pi}\right)^{2k}\right\} . \tag{5.106}$$

The result (5.106) is, as already remarked, strictly U(1) gauge invariant to all orders in the expansion. This excludes the possibility that non-gauge invariant terms such as Chern-Simons terms (see Sec. 7 below for a general discussion of Chern-Simons terms) can arise in the continuum limit and in the exact half-filled case. The same conclusion have been reached by other authors [10–12], who have also concluded that Chern-Simons terms can only arise from non-gauge invariant terms in the Hamiltonian away from half-filling, such as the three-site hopping term of Eq. (2.67).

Most of the analysis of the continuum limit is done under the assumption that $|\mathcal{U}_{ij}|$ is approximately constant, i.e., that:

$$\mathcal{U}_{ij} = \mathcal{U}_0 \exp[i\theta_{ij}] ; \quad \mathcal{U}_0 \simeq \text{const.} \tag{5.107}$$

In view of the transformation law (5.101), the phase θ_{ij} behaves as a normal U(1) lattice gauge variable. The first nontrivial term in the expansion of $\mathcal{F}[\tilde{\mathcal{U}}]$ will come from the last, "plaquette", term of Eq. (5.99), and can be written as:

$$\mathcal{F}_\Box[\tilde{\mathcal{U}}] = \frac{\lambda_2}{\beta}\left(\frac{c}{\pi}\right)^4 \text{Re} \sum_\Box \mathcal{U}_{ij}\mathcal{U}_{jk}\mathcal{U}_{kl}\mathcal{U}_{li} \approx \frac{2\lambda_2}{\beta}\left(\frac{c|\mathcal{U}_0|}{\pi}\right)^2 \sum_\Box \cos(\theta_{ij} + \theta_{jk} + \theta_{kl} + \theta_{li}) \tag{5.108}$$

where the symbol "$\Box$" stands for "plaquette".

Introducing then the "link" gauge field $\mathcal{A}$ as:

$$\exp[i\theta_{ij}] =: \exp\left(i \int_i^j \mathcal{A} \cdot d\mathbf{r}\right) \tag{5.109}$$

one sees at once that the sum of phases on the r.h.s. of (5.108) can be interpreted as the flux $\Phi(\Box)$ through the elementary plaquette of the (fictitious) magnetic field associated with $\mathcal{A}$. This interpretation can be given a more precise meaning in the continuum limit, which we will discuss in more detail in Sec. 6.

To the lowest nontrivial order, $\mathcal{F}_\Box$ contributes to the free energy a term proportional to the square of the flux, and hence to $B^2 =: F_{xy}^2 = (\partial_x \mathcal{A}_y - \partial_y \mathcal{A}_x)^2$. It has been stated in the literature [10–12, 39], that, in the continuum limit, the lattice model becomes a U(1) gauge theory in the axial gauge, and that the extra terms which are needed to set up the full field Lagrangian (which should be proportional to $F_{\mu\nu}F^{\mu\nu}$, $\mu, \nu = x, y, \tau$, and not to B^2 alone) are coming from the "time" dependence of θ_{ij}, and hence ultimately of $\mathcal{A}$, which we have neglected here.

We now discuss briefly how the mean field theories of Affleck and Marston and of Wen, Wilczek and Zee can be rederived from the path-integral approach.

It is known [137] that mean field theory results from a saddle-point evaluation of the path integral. As:

$$\frac{\delta}{\delta\tilde{\mathcal{U}}_{ji}} \mathcal{F}[\tilde{\mathcal{U}}] = \frac{2\pi}{\beta}\left\{\tilde{\mathcal{U}}_{ij} + \frac{c}{2\pi}\mathbb{G}_{ij}(0^-)\right\} \tag{5.110}$$

we see that mean field solutions are the solutions of the nonlinear set of equations:

$$\tilde{\mathcal{U}}_{ij} = -\frac{c}{2\pi}\mathbb{G}_{ij}(0^-) . \tag{5.111}$$

The general structure of the self-consistency equations, as well as the associated stability conditions, have been discussed in Ref. [67], and we refer to the original literature for details. Here we will limit ourselves to the search of minima inside more restricted classes of matrices. In particular, we will consider the effect of imposing on $\tilde{\mathcal{U}}$ (and hence on the Green functions) the condition of translational invariance along the diagonals of a 2D square lattice employed by Affleck and Marston. Our aim is not only to re-obtain the self-consistency Eqs. (5.51), but also to express more generally $\mathcal{F}[\tilde{\mathcal{U}}]$ in closed form for this class of Hubbard-Stratonovich fields.

Dividing, as we did before, the original lattice into two interpenetrating square sublattices, introducing two species of fermions, labeled "c" and "d", living on the two sublattices, and Fourier transforming with respect to the spatial coordinates, we can rewrite the effective Hamiltonian $\mathcal{H}_{\text{eff}}$ in the form (cf. (5.46)):

$$\mathcal{H}_{\text{eff}} = \tilde{J} \sum_{\mathbf{k},\alpha} [\tilde{\lambda}(\mathbf{k}) c^\dagger_{\mathbf{k}\alpha} d_{\mathbf{k}\alpha} + \text{h.c.}] \tag{5.112}$$

where (cf. (5.47)):

$$\tilde{\lambda}(\mathbf{k}) =: \mathcal{U}_1 e^{ik_x a} + \mathcal{U}_2^* e^{-ik_y a} + \mathcal{U}_3 e^{-ik_x a} + \mathcal{U}_4^* e^{ik_y a} \tag{5.113}$$

and the $\mathcal{U}_i$'s, $i = 1, \ldots, 4$, are the four independent values that $\mathcal{U}_{ij}$ can take along the sides of an elementary plaquette.

We are thus led to introduce four Green functions:

$$\mathbb{G}_{cc}(\mathbf{k}; \tau - \tau') = -\sum_\alpha \langle T_\tau [c_{\mathbf{k}\alpha}(\tau) c^\dagger_{\mathbf{k}\alpha}(\tau')]\rangle ;$$

$$\mathbb{G}_{dd}(\mathbf{k}; \tau - \tau') = -\sum_\alpha \langle T_\tau [d_{\mathbf{k}\alpha}(\tau) d^\dagger_{\mathbf{k}\alpha}(\tau')]\rangle \tag{5.114a}$$

and

$$\mathbb{G}_{dc}(\mathbf{k}; \tau - \tau') = -\sum_\alpha \langle T_\tau [d_{\mathbf{k}\alpha}(\tau) c^\dagger_{\mathbf{k}\alpha}(\tau')]\rangle ;$$

$$\mathbb{G}_{cd}(\mathbf{k}; \tau - \tau') = -\langle T_\tau [c_{\mathbf{k}\alpha}(\tau) d^\dagger_{\mathbf{k}\alpha}(\tau')]\rangle . \tag{5.114b}$$

The equations of motion for the Fourier transforms of the Green functions (5.114) can be easily solved [67], to yield:

$$\mathcal{G}_{cc}(\mathbf{k}; \omega_n) = \mathcal{G}_{dd}(\mathbf{k}; \omega_n) ; \quad \mathcal{G}_{dc}(\mathbf{k}; \omega_n) = \mathcal{G}_{cd}^*(\mathbf{k}; \omega_n) \tag{5.115}$$

where:

$$\mathcal{G}_{cc}(\mathbf{k}; \omega_n) = \frac{2i\hbar^2 \omega_n}{(i\hbar\omega_n)^2 - |\tilde{J}\tilde{\lambda}(\mathbf{k})|^2}, \quad (5.116a)$$

$$\mathcal{G}_{dc}(\mathbf{k}; \omega_n) = \frac{2\tilde{J}\hbar\lambda^*(\mathbf{k})}{(i\hbar\omega_n)^2 - |\tilde{J}\tilde{\lambda}(\mathbf{k})|^2}. \quad (5.116b)$$

The poles of the Green function, after analytic continuation from $i\hbar\omega_n$ to the complex variable ζ are then located at:

$$\zeta = \pm E(\mathbf{k}); \quad E(\mathbf{k}) =: \tilde{J}|\tilde{\lambda}(\mathbf{k})| \quad (5.117)$$

that is, the quasiparticle spectrum reproduces (a rescaled version of) the quasiparticle spectrum originally found by Affleck and Marston.

As shown in Ref. [67], the effective free energy $\mathcal{F}[\mathcal{U}]$ can be reconstructed as:

$$\mathcal{F}[\mathcal{U}] = \frac{\pi N}{2\beta} \left\{ \sum_{i=1}^{4} |\mathcal{U}_i|^2 + \frac{4}{\pi N} \sum_{\mathbf{k}} \ln\left(\cosh\left(\frac{\beta E(\mathbf{k})}{2}\right)\right) \right\}. \quad (5.118)$$

Also, the self-consistency equations (5.111) become:

$$\mathcal{U}_1 = -\frac{2\beta \tilde{J}}{\pi N} \sum_{\mathbf{k}} e^{-ik_x a} \frac{\lambda(\mathbf{k})}{2E(\mathbf{k})} \tanh\left(\frac{\beta E(\mathbf{k})}{2}\right), \quad (5.119a)$$

$$\mathcal{U}_2 = -\frac{2\beta \tilde{J}}{\pi N} \sum_{\mathbf{k}} e^{-ik_y a} \frac{\lambda^*(\mathbf{k})}{2E(\mathbf{k})} \tanh\left(\frac{\beta E(\mathbf{k})}{2}\right). \quad (5.119b)$$

The equation for $\mathcal{U}_3$ is the same as (5.119a) with the interchange $k_x \Leftrightarrow -k_x$ in the exponential, and, similarly, the equation for $\mathcal{U}_4$ is obtained from (5.119b) with the interchange $k_y \Leftrightarrow -k_y$ in the exponential.

With the identifications:

$$\chi_i = \frac{\pi J}{c} \mathcal{U}_i. \quad (5.120)$$

Equations (5.119) become identical with (5.51).

The mean-field theory of Wen, Wilczek and Zee [179] can also be rederived very simply in this context. It actually corresponds to adding to the effective Hamiltonian of Eq. (5.112) a tight-binding-type Hamiltonian for each sublattice, generalizing (5.112) into:

$$\mathcal{H}_{\text{eff}} = \tilde{J} \sum_{\mathbf{k}\alpha} [\tilde{\lambda}(\mathbf{k}) c^{\dagger}_{\mathbf{k}\alpha} d_{\mathbf{k}\alpha} + \text{h.c.}] + \tilde{J}' \sum_{\mathbf{k}\alpha} [\epsilon_c(\mathbf{k}) c^{\dagger}_{\mathbf{k}\alpha} c_{\mathbf{k}\alpha} + \epsilon_d(\mathbf{k}) d^{\dagger}_{\mathbf{k}\alpha} d_{\mathbf{k}\alpha}]. \quad (5.121)$$

Indeed [67], one can prove very easily that the quasiparticle energy spectrum is now given by:

$$E_\pm(\mathbf{k}) = \frac{1}{2}\{\tilde{J}'(\epsilon_c + \epsilon_d) \pm \sqrt{\tilde{J}'^2(\epsilon_c - \epsilon_d)^2 + 4\tilde{J}^2|\tilde{\lambda}(\mathbf{k})|^2}\} \quad (5.122)$$

and, with the identifications:

$$\tilde{J}\tilde{\lambda} = \eta_3 \,; \quad \tilde{J}'\epsilon_c = 2\,\mathrm{Re}(\eta_1 + \eta_2)\,; \quad \tilde{J}'\epsilon_d = 2\,\mathrm{Re}(\eta_1 - \eta_2) \quad (5.123)$$

where the η's are defined in Eq. (36) of Ref. [179]. Equation (5.122) becomes identical with Eq. (37) of Ref. [179].

For further details and discussion, we refer the interested reader to the literature [66, 67].

6. Continuum Limit and the Chern-Simons Term

We have seen that the strong-coupling limit of the half-filled Hubbard model is the AFM Heisenberg model. It has been shown [8] that this model has a remarkable SU(2) gauge symmetry. The continuum limit of this model has been studied [23, 193, 195] and a massless Dirac Lagrangian coupled to an SU(2) gauge field has been derived as its approximate long-wavelength description.

It has been argued by Zou [195] that, depending upon the regularization procedure, the effective gauge field action may contain a non-abelian Chern-Simons term. Chern-Simons terms will be extensively discussed in Sec. 7. We will only show here how they can come out in the continuum limit of the Hubbard model.

6.1. *Symmetries of the half-filled Hubbard model in the strong coupling limit*

As we have shown previously, the strong coupling limit of the half-filled Hubbard model is the AFM Heisenberg model whose Hamiltonian can be written as (cf. Eq. (4.10)):

$$\mathcal{H} = -J \sum_{(ij)} b^\dagger_{ij} b_{ij} \qquad (6.1)$$

where the b_{ij}'s have been defined in Eq. (4.6), and one has to implement the half-filling constraint $c^\dagger_{i\alpha} c_{i\alpha} = 1$. It has been shown in Sec. 5 that the model has a U(1) local symmetry: $c_{j\alpha} \to \exp[-i\theta_j] c_{j\alpha}$. We now show that the symmetry is in fact larger, being an SU(2) local symmetry [8].

Introducing for each site the matrix:

$$\psi_i = \begin{bmatrix} c_{i1} & c_{i2} \\ c^\dagger_{i2} & -c^\dagger_{i1} \end{bmatrix}$$

where $c_{i1} = c_{i\uparrow}$ and $c_{i2} = c_{i\downarrow}$, the Heisenberg Hamiltonian can be written as:

$$\mathcal{H} = \frac{J}{16} \sum_{(ij)} [\text{Tr } \psi^\dagger_i \psi_i \sigma^T] \cdot [\text{Tr } \psi^\dagger_j \psi_j \sigma^T] \qquad (6.2a)$$

or in another equivalent form, namely:

$$\mathcal{H} = -\frac{J}{8} \sum_{(ij)} \text{Tr}[\psi_i \psi^\dagger_j \psi_j \psi^\dagger_i] \ . \qquad (6.2b)$$

From Eq. (6.2b) it is immediate that $\mathcal{H}$ is invariant under a global SU(2) spin transformation:

$$\psi_i \to \psi_i g \ ; \quad g \in \text{SU}(2) \ . \qquad (6.3)$$

One can easily see from the form of the Hamiltonian in Eq. (6.2b) that there is a second SU(2) local symmetry present in the model. Under this local SU(2), ψ_i transforms as:

$$\psi_i \to h_i \psi_i ; \quad h_i \in SU(2) . \tag{6.4}$$

Using the form of $\mathcal{H}$ in Eq. (6.2a) it is important to realize that this local symmetry is not a symmetry of the Heisenberg model *per se* [8]. This is because it acts trivially on the spin operators (the global SU(2) spin operators can be written as: $\mathbf{S}_i = [\text{Tr } \psi^\dagger_i \psi_i \sigma^T]/4$, which are invariant under this local SU(2)). It is therefore a symmetry proper of the large U limit of the Hubbard model.

The Lagrangian corresponding to (6.1) is: $\mathcal{L} = \sum_i c_i^\dagger (idc/dt) - \mathcal{H}$. Using the matrices ψ_i it can be written as:

$$\mathcal{L} = \frac{1}{2} \sum_i \text{Tr } \psi^\dagger_i \left(i \frac{d}{dt} \right) \psi_i - \mathcal{H} . \tag{6.5}$$

Although this Lagrangian is invariant under time-independent gauge transformations of the form (6.4), it is not so under time-dependent ones, unlike $\mathcal{H}$. Another problem is that to completely specify the strong coupling limit of the half-filled Hubbard model, one has to impose the constraint of one electron per site:

$$c^\dagger_i c_i - 1 \equiv \frac{1}{2} \text{Tr}[\psi_i^\dagger \sigma^3 \psi_i] = 0 . \tag{6.6}$$

This constraint is invariant under the global SU(2) transformations (6.3), but it is not invariant under the local SU(2) transformations (6.4).

It was realized by Affleck *et al.* [8] that both these problems can be solved together if one introduces Lagrange multiplier fields A_0^a ($a = 1, 2, 3$) which transform as the temporal component of the SU(2) gauge field. The Lagrangian then becomes:

$$\mathcal{L} = \frac{1}{2} \sum_i \text{Tr } \psi^\dagger_i \left(i \frac{d}{dt} + A_{0i} \right) \psi_i - \mathcal{H} \tag{6.7}$$

where A_0 is a traceless, Hermitian matrix: $A_0 = (\sigma^a A_0^a)/2$. Under the local SU(2), A_0 transforms as:

$$A_0 \to h \left[A_0 + i \frac{d}{dt} \right] h^\dagger ; \quad h \in SU(2) . \tag{6.8}$$

The Lagrangian now has full SU(2) gauge invariance. Since the A_0^a are Lagrange multiplier fields, they impose the following constraints:

$$\text{Tr } \psi^\dagger_i \sigma^a \psi_i = 0 ; \quad a = 1, 2, 3 . \tag{6.9}$$

In terms of the original fermion operators these constraints are:

$$c^\dagger_1 c^\dagger_2 = 0 \; ; \quad c_1 c_2 = 0 \quad \text{and} \quad c^\dagger_1 c_1 + c^\dagger_2 c_2 = 1 \tag{6.10}$$

implying, respectively, that: a) there is no vacant site, b) there is no doubly occupied site and c) there is one particle per site. The first two are obviously equivalent to the third one.

As the Lagrangian (6.7) contains only the temporal component of an SU(2) gauge field, it is natural to ask if we can find the spatial components of the gauge field as well. This can be achieved by introducing a Hubbard-Stratonovich field $\mathcal{U}_{ij}$ (not to be confused with the similar field introduced in Sec. 5) given by:

$$\mathcal{U}_{ij} = \frac{J}{8} \psi_i \psi^\dagger_j . \tag{6.11}$$

The Lagrangian (6.7) can then be rewritten as:

$$\mathcal{L} = \frac{1}{2} \sum_i \text{Tr } \psi^\dagger_i \left(i \frac{d}{dt} + A_{0i} \right) \psi_i + \sum_{(ij)} \text{Tr}[\psi^\dagger_i \mathcal{U}_{ij} \psi_j + \text{h.c.}] + \frac{8}{J} \sum_{(ij)} \text{Tr } \mathcal{U}^\dagger_{ij} \mathcal{U}_{ij} . \tag{6.12}$$

From (6.11) we can deduce the transformation properties of $\mathcal{U}_{ij}$ under the SU(2) gauge transformations, namely:

$$\mathcal{U}_{ij} \to h_i \mathcal{U}_{ij} h^\dagger_j . \tag{6.13}$$

Note that $\mathcal{U}_{ij}$ transforms in the same way as the lattice variables in a SU(2) lattice gauge theory (cf. the similar discussion in Sec. 5, where only the U(1) gauge symmetry was taken into account). However, in the case of a lattice group theory, the matrix $\mathcal{U}$ is restricted to be an SU(2) matrix. To make this correspondence more precise, one can write $\mathcal{U}_{ij}$ as:

$$\mathcal{U}_{ij} = \mathcal{U}_0 \hat{\mathcal{U}}_{ij}$$

where $\hat{\mathcal{U}}_{ij}$ is an SU(2) matrix.

For low energies, one expects that the integral over $\mathcal{U}_0$ will be dominated by a nonzero saddle point with fluctuations being unimportant [8, 193]. Thus, the low energy sector of the model will be described by normal lattice theory variables.

The spatial component of the vector potential can now be defined by writing:

$$\hat{\mathcal{U}}_{ij} = \mathcal{P} \left\{ \exp \left[i \int_i^j dl A(l) \right] \right\} \tag{6.14}$$

where $A(l)$ lives on the link joining two neighboring sites i and j. A is a Hermitean, traceless matrix given by:

$$A = \sigma^a A^a .$$

This completes our specification of the model with an SU(2) global spin symmetry as well as full SU(2) gauge invariance.

6.2. The continuum limit

The continuum limit of the strongly coupled half-filled Hubbard model was derived in Ref. [193]. It was shown that the model reduces, in the continuum limit, to a massless Dirac Lagrangian coupled to an SU(2) gauge field. It was shown in Ref. [23] that essentially the same continuum limit is obtained by considering a class of continuum models which possess all the desired symmetries present in the lattice model.

Although the SU(2) gauge symmetry of the lattice model is manifestly present in the Dirac Lagrangian coupled to the SU(2) gauge field at the continuum level, this is not obvious with the SU(2) global spin symmetry of the lattice model. We would like to first illustrate this point in the following before getting on to the derivation of the continuum limit.

Symmetries of the Dirac Lagrangian

It is well known that for taking the continuum limit one must introduce two separated fermion fields living on even and odd lattice sites respectively. The reason for this is essentially the following [119]. The spatial derivatives in the continuum model arise from terms of the kind $\psi^\dagger_n \psi_m$ in the lattice model. Upon transformation to the continuum variables this term leads to the term $\psi^\dagger_n (\psi_{n+1} - \psi_{n-1})$ (for nearest neighbor coupling). Here the relative sign appears due to the presence of factors i^n in the transformation equations (see Eq. (6.29) below). Thus, spatial derivatives are defined in terms of $(\psi_{n+1} - \psi_{n-1})/2a$, where a is the lattice spacing ($a \to 0$ in the continuum limit). The requirement that the derivatives remain finite imposes the restriction that $(\psi_{n+1} - \psi_{n-1}) \to 0$ as $a \to 0$.

However, since terms of the kind $\psi^\dagger_n (\psi_{n+1} - \psi_n)$ do not appear in going to the continuum limit, $\psi_{n+1} - \psi_n$ is not constrained to vanish as $a \to 0$.

Thus in order to have fields with finite derivatives one is led to define two separate fields for even and odd lattice sites (a lattice site $n \equiv (n_x, n_y)$ is called even (odd) if $n_x + n_y$ is even (odd), where n_x, n_y represent lattice displacements in the x and y directions respectively).

Returning back to our problem, we thus introduce two fields living at even (odd) lattice sites respectively, for the continuum limit of Eq. (6.12). This leads to consideration of the following matrix [23, 193]:

$$\mathscr{C}_M = \begin{bmatrix} c_1^{(e)} & c_2^{(e)} \\ c_2^{(e)\dagger} & -c_1^{(e)\dagger} \\ c_1^{(o)} & c_2^{(o)} \\ c_2^{(o)\dagger} & -c_1^{(o)\dagger} \end{bmatrix} \qquad (6.15)$$

where e and o denote even and odd lattice sites respectively.

If we make the definitions:

$$\Psi = (\Psi^{(\alpha)}), \qquad \rho, \alpha = 1, 2$$

$$(\Psi_1^{(1)}, \Psi_1^{(2)}) = (c_1^{(e)}, c_2^{(e)\dagger}) \tag{6.16}$$

$$(\Psi_2^{(1)}, \Psi_2^{(2)}) = (c_1^{(o)}, c_2^{(o)\dagger})$$

the Dirac field in the continuum model will be associated with $\Psi_\rho^{(\alpha)}$, or the first column of $\mathscr{C}_M$ [23, 193]. The Lorentz group symmetry of the Dirac Lagrangian will be associated with the action of that group on the index ρ of $\Psi_\rho^{(\alpha)}$. The SU(2) spin symmetry (6.3) then suggests that this SU(2) action must mix $\Psi_\rho^{(\alpha)}$ and its adjoint. The existence of such an SU(2) action leaving the Dirac Lagrangian unaltered is not however obvious. Let us demonstrate the existence of this symmetry first by considering a massless Dirac Lagrangian without gauge invariance. We will later incorporate gauge invariance.

The two-dimensional irreducible representation of the 2 + 1 Lorentz group is isomorphic to SL(2, $\mathbb{R}$) and hence can be realized by real matrices. It is particularly convenient for us to use this realization in our discussion. With this in mind, we choose the following γ-matrices:

$$\gamma^1 = \sigma_3, \qquad \gamma^2 = \sigma_1, \qquad \gamma^3 = i\gamma^0 = \sigma_2, \tag{6.17}$$

$$\sigma_i = \text{Pauli matrices}.$$

With this identification, the generators

$$\frac{i}{4}[\gamma^\mu, \gamma^\nu]; \qquad \mu, \nu = 0, 1, 2$$

of the Lorentz group Lie algebra are imaginery just as we want.

The massless Dirac Lagrangian is

$$\mathscr{L}_0 = \int d^2x \bar{\psi}(x) \gamma^\lambda \partial_\lambda \psi(x) \tag{6.18}$$

$$\bar{\psi} = \psi^\dagger \gamma^3; \qquad x = x^0, x^1, x^2; \qquad \partial_\lambda = \frac{\partial}{\partial x^\lambda}$$

where ψ is a two-component spinor (corresponding to $\Psi_\rho^{(\alpha)}$ for fixed α). The Lorentz group acts on ψ according to:

$$\psi(x) \to s\psi[\Lambda(s)^{-1}x]; \qquad s \in \text{SL}(2, \mathbb{R}) \tag{6.19}$$

where $\Lambda(s)$ is defined by

$$s\gamma_\mu s^{-1} = \gamma_\nu \Lambda(s)^\nu_{\ \mu}. \tag{6.20}$$

Since s is real, we see that ψ and ψ^* transform in the same way under the Lorentz group. A consequence of this fact is that $\mathcal{L}_0$ is invariant under the SU(2) group acting on ψ and ψ^*. Such a result was proved long ago by Pauli and Gürsey [87, 145, 146] in 3+1 dimensions. We can bring out this invariance of $\mathcal{L}_0$ explicitly by introducing the matrix of fields:

$$\chi = \begin{bmatrix} \psi_1 & \psi_1^* \\ \psi_2 & \psi_2^* \end{bmatrix}. \quad (6.21)$$

Then $SL(2, \mathbb{R})$ acts according to

$$\chi(x) \to s\chi(\Lambda(s)^{-1}x)$$

whereas SU(2) acts by right multipliation:

$$\chi(x) \to \chi(x)h^T; \quad h \in SU(2). \quad (6.22)$$

Now (6.18) can be written as follows on using the Grassmann nature of the Dirac field:

$$\mathcal{L}_0 = \frac{1}{2}\int d^2x \, \mathrm{Tr}\, \bar{\chi}(x)\gamma^\lambda \partial_\lambda \chi(x),$$
$$\bar{\chi} = \chi^\dagger \gamma^3. \quad (6.23)$$

The Lorentz and SU(2) symmetries of $\mathcal{L}_0$ are manifest in (6.23).

Let us now promote ψ to a two-component field $\psi^{(\alpha)}$ ($\alpha = 1, 2$) and consider

$$\mathcal{L}_1 = \sum_{\alpha=1}^{2}\int d^2x \bar{\psi}^{(\alpha)}(x)\gamma^\lambda \partial_\lambda \psi^{(\alpha)}(x). \quad (6.24)$$

The Lagrangian in this case has a new global SU(2) symmetry:

$$\psi^{(\alpha)} \to g_{\alpha\beta}\psi^{(\beta)}, \quad g \in SU(2). \quad (6.25)$$

$\psi^{(\alpha)*}$ does not transform in the same way as $\psi^{(\alpha)}$ under this group. However, $(i\tau_2\psi^*)^{(\alpha)}$ and $\psi^{(\alpha)}$ do transform in the same way where

$$(i\tau_2\psi^*)^{(\alpha)} = (i\tau_2)_{\alpha\beta}\psi^{(\beta)*}$$

and τ_i are Pauli matrices. The appropriate generalization of χ is thus:

$$\chi^{(\alpha)} = \begin{bmatrix} \psi_1^{(\alpha)} & (i\tau_2\psi_1^*)^{(\alpha)} \\ \psi_2^{(\alpha)} & (i\tau_2\psi_2^*)^{(\alpha)} \end{bmatrix}. \quad (6.26)$$

In terms of this field, (6.24) becomes

$$\mathcal{L}_1 = \frac{1}{2} \sum_{\alpha=1}^{2} \int d^2x \, \text{Tr} \, \bar{\chi}^{(\alpha)}(x) \gamma^\lambda \partial_\lambda \chi^{(\alpha)}(x) \, . \tag{6.27}$$

The Lorentz and Pauli-Gürsey groups, and the SU(2) group of (6.25), act on $\chi^{(\alpha)}$ in an obvious way. The invariance of $\mathcal{L}_1$ under these actions is manifest in (6.27).

The "internal" SU(2) symmetry (6.25) can now be gauged by replacing ∂_λ by $D_\lambda = \partial_\lambda + iA_\lambda$, A_λ being the SU(2) connection field. On doing so, $\mathcal{L}_1$ becomes

$$\mathcal{L}_2 = \frac{1}{2} \int d^2x \, \text{Tr} \, \bar{\chi} \gamma^\lambda D_\lambda \chi \tag{6.28}$$

where we have suppressed the index α.

This Lagrangian is invariant under the actions of the Pauli-Gürsey group, the SU(2) gauge group and the 2 + 1 dimensional Lorentz group.

Being certain of the symmetries of the lattice model correctly represented in (6.28) (though the spatial symmetries of (6.28) are larger than the ones in the lattice model. For a discussion of this, see Ref. [23]) let us now discuss the deviation of (6.28) from the lattice model.

Derivation of the continuum limit

As we have mentioned earlier, the continuum limit of (6.12) is a massless Dirac field coupled to an SU(2) gauge field. This can be shown by deriving the continuum model from the lattice model by letting the lattice constant $a \to 0$ and retaining terms linear in a [193]. The same continuum limit is obtained by other considerations, as shown in Refs. [195] and [23].

Following Ref. [193] we shall present in the following a derivation of the continuum limit for a 1+1 dimensional system. This will illustrate the techniques and approximations involved in taking such a limit (see also Kogut and Susskind in Ref. (119)). The derivation for the 2+1 dimensional system can then be carried out in a similar manner, and we shall simply present the result for that case (see Refs. [23, 193, 195]).

In the continuum limit, the lattice constant $a \to 0$ and one can expand $\hat{U}_{ij}$ in (6.14) as

$$\hat{U}_{i,i+1} = \exp[iAa] \approx 1 + iAa + O(a^2) \, .$$

As we have mentioned, in the continuum limit one must introduce two separated fermion species, say ψ_o and ψ_e which live on odd and even lattice sites respectively. The relation between the continuum variables ψ_o and ψ_e and the lattice variables ψ_i is the following:

$$\psi_{2n} \to i^{2n}(2a)^{1/2}\psi_e(x, t) \, ,$$
$$\psi_{2n+1} \to i^{2n+1}(2a)^{1/2}\psi_o(x, t) \, . \tag{6.29}$$

Then the spatial derivative of ψ_o is defined as:

$$\partial_x \psi_o(x, t) = \lim_{a \to 0} \frac{\psi_{2n+1} - \psi_{2n-1}}{2a}. \qquad (6.30)$$

Similarly, $\partial_x \psi_e$ can also be defined. The action for (6.12), neglecting the last term, can then be obtained by replacing the summation over lattice sites by an integral. We get:

$$S = \frac{1}{2}\int dxdt \ \text{Tr}\{\psi_e^\dagger D_t \psi_e + \psi_o^\dagger D_t \psi_o\} + \frac{i}{2}\int dxdt \ \text{Tr}\{\psi_e^\dagger D_x \psi_o + \psi_o^\dagger D_x \psi_e\} \qquad (6.31)$$

where we have set $U_0 a = \frac{1}{4}$ and $D_\mu = \partial_\mu + iA\mu$ is the covariant derivative. By defining

$$\psi_{r(l)} = \frac{\psi_e \pm \psi_o}{\sqrt{2}}$$

we can write it as

$$S = \frac{1}{2}\int dxdt \{\text{Tr}[\psi_r^\dagger D_t \psi_r + \psi_l^\dagger D_t \psi_l] + i \ \text{Tr}[\psi_r^\dagger D_x \psi_r - \psi_l^\dagger D_x \psi_l]\} . \qquad (6.32)$$

Let us introduce the chiral field

$$\chi_{l(r)} = \begin{bmatrix} c_{1l(r)} \\ c_{2l(r)}^\dagger \end{bmatrix}$$

and the chiral doublet

$$\chi = \begin{bmatrix} \chi_r \\ \chi_l \end{bmatrix}$$

(where l, r subscripts are again defined in terms of even and odd lattice sites as done above in Eq. (6.32)).

Then it is easy to see that

$$\int dxdt \ \text{Tr} \ \psi_r^\dagger D_\mu \psi_r = 2 \int dxdt \chi_r^\dagger D_\mu \chi_r , \text{ etc.}$$

The action in (6.32) can then be written as

$$S = \int d^2x \bar{\chi} \gamma^\mu D_\mu \chi \qquad (6.33)$$

where γ_0, $\gamma_1 = \sigma_1$, σ_2 and $\bar{\chi} = \chi^\dagger \gamma_0$.

Generalization of the above procedure to the 2+1 dimensional case is rather straightforward, the result being again a massless Dirac Lagrangian in 2+1 dimensions [23, 193, 195]. After considering the doubling of fermion species (which is expected in going to the continuum limit [119, 143, 144]. However, see the comments in Ref. [23] regarding the question of doubling of species for a 2+1 dimensional system) the continuum model can be written as [195]:

$$S = \sum_{a=1}^{2} \int d^3x \bar{\Psi}_a \gamma^\mu D_\mu \Psi_a$$

$$\Psi_a = \begin{bmatrix} \psi_e \\ \psi_o \end{bmatrix}; \quad \psi_{e(o)} = \begin{bmatrix} c_{1e(o)} \\ c_{2e(o)}^\dagger \end{bmatrix}$$

(6.34)

and the γ matrices are given as in Eq. (6.17). (Note that our choice of the γ matrices is different from those in Refs. [193, 195].) Our choice was governed by our considerations of the SU(2) spin symmetry in the Dirac Lagrangian. (6.34) is the gauged version of (6.23), and hence can be written as (6.28), which is manifestly invariant under the SU(2) gauge group and the Lorentz group.

6.3. *The Chern-Simons term*

The effective action for the gauge field can now be obtained from (6.34) by integrating out the fermion degrees of freedom. It was shown by Redlich [152, 153] for the case of one species of fermions that the effective action

$$S_{\text{eff}}(A) = \ln \text{Det}[\gamma^\mu(\partial_\mu + iA\mu)]$$

is not invariant under a large gauge transformation. The meaning of a "large" gauge transformation can be understood as follows. In an Euclidean formulation of the 2+1 dimensional gauge theory one imposes the restriction on the gauge transformations that they approach identity at large distances. This leads to the consideration of gauge transformations on a manifold which is S^3 rather than $\mathbb{R}^3$. These gauge transformations for the gauge group SU(2) (in fact for SU(N) for any $N > 1$) fall into homotopy classes which are simply characterized by winding numbers of one sphere S^3 (which is the base manifold) onto another sphere S^3 (the group manifold of SU(2)). A gauge transformation which corresponds to a nonzero winding number is called a "large" gauge transformation. "Small" gauge transformations correspond to zero winding number.

Redlich showed that under a large gauge transformation g_n with winding number n:

$$g_n: S_{\text{eff}}(A) \to S_{\text{eff}}(A) \pm \pi|n| .$$

The gauge symmetry can be restored by a gauge invariant Pauli-Villars regularization. However, $S_{\text{eff}}(A)$ then contains a parity violating topological term which is the Chern-Simons term:

$$S_{CS} = \frac{1}{8\pi^2} \int d^3x \, \text{Tr} \, \epsilon_{\mu\nu\lambda} \left[\frac{1}{2} A^\mu F^{\nu\lambda} - \frac{1}{3} A^\mu A^\nu A^\lambda \right].$$

The sign of S_{CS} in the effective action depends upon the sign of the mass of the fermion introduced to regularize the fermion determinant [152, 153].

Since, in our case, (6.34) describes two species of fermions, one may introduce a parity preserving mass term for the two fermions in order to regularize the fermion determinant. Thus consider

$$\sum_{a=1}^{2} \bar{\Psi}_a \gamma^\mu D_\mu \Psi_a + m(\bar{\Psi}_1 \Psi_1 - \bar{\Psi}_2 \Psi_2) \,. \tag{6.35}$$

One can then immediately see that the Chern-Simons terms from the two species of fermions would have opposite signs and they will cancel each other. Thus, in this regularization scheme one does not get any Chern-Simons term in the effective action.

However, it was argued in [195] that, instead of (6.35), one may consider the Pauli-Villars regulator with the same sign for the two fermion mass terms. One then gets the Chern-Simons term with twice the coefficient in the effective action $S_{\text{eff}}(A)$. It was further argued in [195] that because of this coefficient 2, the quasiparticle excitations will be half-fermions. For a discussion of the Chern-Simons term and its relation to statistics, see Sec. 7.

Following a different line of approach and considering only the U(1) gauge symmetry of the Heisenberg model (see Sec. 6.1 and the end of Sec. 5) Aitchison and Mavromatos [10–12] argued that a U(1) topological Chern-Simons term

$$\frac{i}{4\pi} \int d^3x \, \epsilon^{\mu\nu\lambda} A_\mu \partial_\nu A_\lambda \tag{6.36}$$

will be introduced when considering the continuum limit of the doped antiferromagnetic Heisenberg model. The approach followed in [10–12] uses the effective field theory method of Baskaran and Anderson [39]. By introducing effective degrees of freedom Δ_{ij}, to be regarded as variables linking sites i and j in the sense of lattice gauge theory, such that $\langle \Delta_{ij} \rangle = \langle b_{ij} \rangle$ (where the b_{ij}'s have been given in Eq. (4.6)), it was shown in [10–12] that the continuum limit is a U(1) gauge theory action.

By arguing that the doping induces a term (actually coming from the three-site hopping term discussed in Sec. 2 (Eq. (2.67))

$$d \sum_{(ijk)} (\Delta_{ij}^* \Delta_{jk} + \text{h.c.}) \tag{6.37}$$

in the effective action, with $d \propto \delta$ (the doping fraction) and with the summation restrictions being the same as in the second sum on the r.h.s. of Eq. (5.99), it was argued in [10–12] that the Chern-Simons term (6.36) arises from (6.37) in the continuum limit. We will not discuss this model further here, and refer the reader to [10–12] for details.

7. The Abelian Chern-Simons Term

In odd space-time dimensions, the Lagrangian of any system involving the Yang-Mills potential of an Abelian or a semi-simple Lie group can be augmented by a remarkable topological term [30, 31, 57, 58, 62, 76, 79, 83–86, 90, 112, 148, 155, 158, 161, 163, 164, 175, 184, 191]. It is called the Chern-Simons (CS) term. Such a term exists also for models involving fields valued in certain coset spaces [13, 26, 32, 46, 59–61, 63, 71, 128, 147, 159, 168, 169, 177, 185, 189] or gauge groups with Lie algebras which admit an invariant bilinear form [186, 187]. In this section we briefly review the construction and properties of the abelian version of this term both for U(1) gauge theories and for $\mathbb{C}P^1$ models in 2+1 dimensions. The nonabelian version of this term will be treated in Sec. 8. Recent studies suggest that Chern-Simons terms may be important both for an understanding of the low energy sector of the Hubbard model and for the description of a novel class of two-dimensional superconductors. In this section, we limit the discussion to a review of the basic properties of the 2+1 dimensional abelian Chern-Simons term, and shall return in later sections to the treatment of these terms in the context of two-dimensional condensed matter systems.

7.1. *U(1) Gauge theories*

The Chern-Simons term in the Lagrangian density for a U(1) gauge theory with the connection field $A = A_\mu dx^\mu$ is

$$\mathscr{L}_{CS} = \frac{k}{4\pi} \epsilon^{\mu\nu\lambda} A_\mu \partial_\nu A_\lambda . \tag{7.1}$$

Although $\mathscr{L}_{CS}$ is not invariant under the gauge transformation

$$A_\mu \to A_\mu + \partial_\mu \Lambda , \tag{7.2}$$

it changes only by a divergence:

$$\mathscr{L}_{CS} \to \mathscr{L}_{CS} + \frac{k}{4\pi} \partial_\mu \epsilon^{\mu\nu\lambda} \Lambda \partial_\nu A_\lambda . \tag{7.3}$$

Consequently the equations of motion are gauge invariant and the transformation (7.2) must be regarded as defining the gauge group of the theory.

There is a useful way to understand the simple response of $\mathscr{L}_{CS}$ to the gauge transformation (7.2). If

$$F_{\mu\nu} = \partial_\mu A_\nu - \partial_\nu A_\mu , \tag{7.4}$$

is the field strength, let us introduce the curvature two-form

$$F = F_{\mu\nu} dx^\mu \wedge dx^\nu . \tag{7.5}$$

Now $F_{\mu\nu}$ satisfies the Bianchi identity

$$\partial_\mu F_{\nu\lambda} + \partial_\nu F_{\lambda\mu} + \partial_\lambda F_{\mu\nu} = \frac{1}{2} \epsilon^{\mu\nu\lambda} \partial_\mu F_{\nu\lambda} = 0 , \tag{7.6}$$

which means that F is closed:

$$dF = 0 . \tag{7.7}$$

Next consider the four-form

$$\omega^{(4)} = \frac{k}{16\pi} F \wedge F . \tag{7.8}$$

It is closed because F is closed, regardless of space-time dimensions:

$$d\omega^{(4)} = 0 . \tag{7.9}$$

Hence by the Poincaré lemma, there is a three-form ω_{CS} such that

$$\omega^{(4)} = d\omega_{\text{CS}} . \tag{7.10}$$

The components of this three-form can be associated with $\mathcal{L}_{\text{CS}}$ since a simple calculation shows that in 2+1 dimensions

$$d[\mathcal{L}_{\text{CS}} d^3 x] = \omega^{(4)} . \tag{7.11}$$

The general solution of (7.10) is not of course $\mathcal{L}_{\text{CS}} d^3 x$, but rather

$$\omega_{\text{CS}} = \mathcal{L}_{\text{CS}} d^3 x + d\Phi , \tag{7.12}$$

Φ being any two-form.

Now $\omega^{(4)}$ being gauge invariant, the change in any particular solution ω_{CS} under (7.2) must be by a term of the kind $d\Phi$. Translated in components, this means that $\mathcal{L}_{\text{CS}}$ must transform by a total divergence as in (7.3).

We shall see that this way of thinking about the transformation properties of the Chern-Simons term is very effective especially when discussing its non-Abelian generalization.

The presence of $\mathcal{L}_{\text{CS}}$ in the Lagrangian density of a dynamical system has many remarkable consequences. Here we shall review certain of these consequences which are of particular interest in condensed matter physics.

7.2. *Mass generation, Meissner effect, London equation*

Let us consider a field theory with the Lagrangian density

$$\mathcal{L} = -\frac{1}{4} F_{\mu\nu} F^{\mu\nu} + \mathcal{L}_{CS} . \tag{7.13}$$

It is thus the standard Lagrangian density for a U(1) gauge field augmented by the topological term $\mathcal{L}_{CS}$. The gauge field quanta described by $\mathcal{L}$ are massless in the limit $k \to 0$ which sends the topological term to zero. It is a remarkable fact that this is no longer the case if $k \neq 0$ as we shall now show.

The field equation for $\mathcal{L}$ is

$$\partial_\nu F^{\mu\nu} = -\frac{k}{4\pi} \epsilon^{\nu\lambda\rho} F_{\lambda\rho} . \tag{7.14}$$

Multiplying by $\epsilon^{\alpha\beta\nu}$, applying ∂_α, and using the Bianchi identity $\epsilon^{\mu\nu\lambda} \partial_\mu F_{\nu\lambda} = 0$, we get

$$-\Box \epsilon^{\beta\nu\alpha} F_{\nu\alpha} + \epsilon^{\beta\nu\alpha} \partial_\nu \partial_\mu F_{\mu\alpha} = -\frac{k}{4\pi} \epsilon^{\alpha\beta\nu} \partial_\alpha \epsilon^{\nu\lambda\rho} F_{\lambda\rho} . \tag{7.15}$$

In view of (7.14), this means

$$\Box *F_\beta = \frac{k}{\pi} \partial^\alpha F_{\alpha\beta} = -\frac{k^2}{4\pi^2} *F_\beta , \tag{7.16}$$

where $*F_\beta = \frac{1}{2} \epsilon_{\beta\gamma\alpha} F^{\gamma\alpha}$ is the dual to the tensor $F^{\gamma\alpha}$, which shows clearly that the gauge invariant field strength F is a free field of mass $k^2/4\pi^2$. This is the Meissner effect. The gauge invariant form of the London equation for superconductivity.

$$\partial_\mu J_\nu - \partial_\nu J_\mu = F_{\mu\nu} , \tag{7.17}$$

connecting the current J_μ $(:= \partial^\lambda F_{\lambda\mu} = -(k^2/4\pi^2)\epsilon_{\mu\nu\lambda} F^{\nu\lambda})$ and the field strength is also valid here as a simple manipulation shows.

7.3. *Fractional statistics*

The possible statistics of N identical particles moving in $\mathbb{R}^k$ ($K \geq 3$) are characterized by the unitary irreducible representations (UIR's) of the permutation group S_n [29]. The group S_n is generated by transpositions or exchanges σ_i. The physical significance of σ_i is that, acting on an N particle state, it exchanges the coordinates of the particles i and $i + 1$. It is known that S_n has the following presentation in terms of exchanges:

$$S_n = \langle \sigma_1, \sigma_2, \ldots, \sigma_{n-1} | \sigma_i^2 = e;$$
$$\sigma_i \sigma_{i+1} \sigma_i = \sigma_{i+1} \sigma_i \sigma_{i+1}; \sigma_i \sigma_j = \sigma_j \sigma_i \quad \text{for } |i - j| \geq 2 \rangle , \tag{7.18}$$

where e is the identity of S_n. When the states of N identical particles are invariant under exchanges and hence transform trivially under S_n, the particles are bosons, whereas if these states change sign under exchanges, they are fermions. Both these UIR's are one-dimensional. S_n for $N \geq 3$ also have UIR's of dimension larger than 1. Particles associated with these higher dimensional UIR's are called paraparticles. They are called parabosons or parafermions depending on the particular UIR's which characterize their multiparticle states.

The situation with regard to statistics is quite different when $K = 2$ [29, 45, 73, 74]. When $K = 2$, it can be shown that statistics is governed not by the group S_n, but by a group known as the braid group B_n. It has the presentation

$$B_n = \langle \sigma_1, \sigma_2, \ldots, \sigma_{n-1} | \sigma_i \sigma_{i+1} \sigma_i = \sigma_{i+1} \sigma_i \sigma_{i+1}; \sigma_i \sigma_j = \sigma_j \sigma_i \text{ for } |i - j| \geq 2 \rangle, \tag{7.19}$$

where σ_i again has the significance of the exchange of particles i and $i + 1$. It is thus obtained from (7.18) by deleting the condition $\sigma_i^2 = e$. The consequences of this deletion are serious. Thus, for instance, B_n is an infinite group and its every element except the identity is of infinite order.

Since the elements of S_n are of finite order, it follows from latter observation that S_n is not a subgroup of B_n.

It is however true that S_n is a factor group of B_n factored by a normal subgroup P_n. The subgroup P_n is known as the pure braid group and is the smallest normal subgroup of B_n containing σ_i^2 for all i. Therefore, every representation of S_n can be found among those of B_n.

There is a class of UIR's of B_n, consisting of all its one-dimensional UIR's, which are of particular interest for the current discussion. They are obtained by representing σ_i by the phases $\exp[i\theta_i]$ (θ_i real). The relation $\sigma_i \sigma_{i+1} \sigma_i = \sigma_{i+1} \sigma_i \sigma_{i+1}$ then shows that $\exp[i\theta_i] = \exp[i\theta_j]$ for all i, j. Calling $\exp[i\theta_i]$ as $\exp[i\theta]$, we thus see that the states acquire the phase $\exp[i\theta]$ under particle exchange for this UIR. The particles associated with this UIR are said to obey fractional statistics. For $\theta = 0, \pi \pmod{2\pi}$, they become bosons and fermions. Hence in two dimensions, fractional statistics interpolates between Bose and Fermi statistics.

The parameter $k/4\pi$ in the Chern-Simons term $\mathcal{L}_{CS}$ is related to the angle θ which describes fractional statistics. This comes about as follows. Let us consider N identical particles of charge e in interaction with the abelian potential A_μ. The particle Lagrangian is:

$$L_\text{particle} = \frac{1}{2} \sum_\alpha \dot{\mathbf{z}}_{(\alpha)}^2(t) + e \sum_\alpha A_\mu[\mathbf{z}_{(\alpha)}(t), t] \dot{z}_{(\alpha)}{}^\mu$$

$$z_{(\alpha)}{}^0(t) = t \tag{7.20}$$

whereas for the field, we will consider the Lagrangian density to be the Chern-Simons term:

$$\mathscr{L}_{\text{field}} = \mathscr{L}_{\text{CS}} . \tag{7.21}$$

The field Lagrangian will in general contain many more terms such as constant $\times F_{\mu\nu}F^{\mu\nu}$ but for large distances and low energies, the replacement of a field Lagrangian by $\mathscr{L}_{\text{CS}}$ should be a good approximation (if $k \neq 0$). This is because it contains fewer derivatives than terms such as constant $\times F_{\mu\nu}F^{\mu\nu}$.

The field equations which follow from (7.20) and (7.1) are

$$\frac{k}{2\pi} F_{\mu\nu} = -\epsilon_{\mu\nu\lambda} e \sum_\alpha \int dt \delta^3[\mathbf{x} - \mathbf{z}(t)] \dot{z}_{(\alpha)}{}^\lambda(t) \tag{7.22}$$

whereas the equations of motion for the particles are

$$m\ddot{z}_{(\alpha)i}(t) = eF_{i\lambda} \dot{z}_{(\alpha)}{}^\lambda(t) . \tag{7.23}$$

It is interesting and important to note that the particle motions in classical theory are actually free. For on inserting (7.22) in (7.23) we find

$$\ddot{\mathbf{z}}_{(\alpha)}(t) = 0 . \tag{7.24}$$

The particle will in general scatter in quantum mechanics however, so that they can then no longer be regarded as free [9, 13, 59, 60, 89].

Following Arovas et al. [27], we shall now derive a simple effective Lagrangian involving only the particle coodinates and infer the statistics of the sources from this Lagrangian. It is constructed by first solving (7.22) for A_μ. The effective interaction between the particles is then

$$e \sum_\alpha A_\mu[\mathbf{z}_{(\alpha)}] \dot{z}_{(\alpha)}{}^\mu + \int d^3x \mathscr{L}_{\text{CS}} = \frac{1}{2} \sum_\alpha A_\mu[\mathbf{z}_{(\alpha)}] \dot{z}_{(\alpha)}{}^\mu , \tag{7.25}$$

where we have used (7.22) to simplify the second term. For the moment we shall neglect the self-interaction of the particles. However we will come back to this interaction later. This means that $A_\mu[\mathbf{z}_{(\alpha)}]$ in (7.25) will be the solution of (7.22) with all sources β excluding α.

The general solution of (7.22) up to a gauge transformation is readily found. It is a superposition of what may be called Aharonov-Bohm potentials and may be described as follows. At time t, let us set up Cartesian coordinate systems $S_{(\alpha)}(t)$ with identical orientations with $\mathbf{z}_{(\alpha)}(t) = (z_{(\alpha)}{}^1(t), z_{(\alpha)}{}^2(t))$ defining their origins. It is convenient to choose $S_{(\alpha)}$ so that they are related to each other by translations. Let $\phi_{(\alpha)}$ be the azimuthal angle of $\mathbf{x} = (x^1, x^2)$ in the coordinate system $S_{(\alpha)}(t)$.

Since $S_{(\alpha)}(t)$ can be permuted by translations, it is clear that $\phi_{(\alpha)}$ is a function of $\mathbf{x} - \mathbf{z}_{(\alpha)}(t)$. So let us denote this angle by $\phi[\mathbf{x} - \mathbf{z}_{(\alpha)}(t)]$. The solution A_μ for all sources β excluding α is then

$$A_\mu(\mathbf{x}) = -\frac{e}{k} \sum_{\beta \neq \alpha} \partial_\mu \phi[\mathbf{x} - \mathbf{z}_{(\beta)}(t)] \tag{7.26}$$

where $x^0 = z_{(\beta)}{}^0(t) = t$.

Let us quickly verify that (7.26) solves (7.22). Let $C_{(\alpha)}$ be a positively oriented closed contour enclosing only the αth source. Since

$$\int_{C_{(\alpha)}} \partial_a \phi[\mathbf{x} - \mathbf{z}_{(\alpha)}(t)] dx^a = 2\pi \tag{7.27}$$

$$\partial_a := \frac{\partial}{\partial x^a} \quad (a = 1, 2),$$

we have, by Stokes' theorem:

$$(\partial_a \partial_b - \partial_b \partial_a)\phi[\mathbf{x} - \mathbf{z}_{(\alpha)}(t)] = 2\pi\delta^2[\mathbf{x} - \mathbf{z}_{(\alpha)}(t)]\epsilon_{ab} \tag{7.28}$$

where the left-hand side is to be understood in the distribution sense. Using this identity, we see easily that (7.26) solves (7.22).

The effective interaction Lagrangian for the particles can now be found from (7.20) and (7.25). It is

$$L_{\text{eff}} = \frac{1}{2} m[\dot{\mathbf{z}}_{(\alpha)}(t)^2] + \frac{1}{2}\left(-\frac{e^2}{k}\right)\sum_{\alpha \neq \beta} \frac{d\phi_{\alpha\beta}}{dt}(t). \tag{7.29}$$

$$\phi_{\alpha\beta}(t) = \phi[\mathbf{z}_{(\alpha)}(t) - \mathbf{z}_{(\beta)}(t)]. \tag{7.30}$$

It shows explicitly that the particle motion is classically free. As emphasized earlier, this does not preclude the possibility of quantum mechanical scattering.

The contribution of the interaction and Chern-Simons terms to L_{eff} are

$$-\frac{e^2}{k}\sum_{\alpha \neq \beta}\dot{\phi}_{\alpha\beta} \quad \text{and} \quad \frac{e^2}{2k}\sum_{\alpha \neq \beta}\dot{\phi}_{\alpha\beta}.$$

This explains the overall factor of $1/2$ in the second term of L_{eff}.

The momentum $p_{(\alpha)a}$ conjugate to $z_{(\alpha)a}$ for L_{eff} is

$$p_{(\alpha)a} = m\dot{z}_{(\alpha)a} - \frac{e^2}{k}\sum_{\beta \neq \alpha}\frac{\partial \phi_{\alpha\beta}}{\partial z_{(\alpha)a}} \tag{7.31}$$

and the Hamiltonian is

$$H_\theta = \frac{1}{2m}\sum_{\alpha,a}\left[p_{(\alpha)a} + \frac{\theta}{\pi}\sum_{\beta \neq \alpha}\frac{\partial \phi_{\alpha\beta}}{\partial z_{(\alpha)a}}\right]^2, \tag{7.32}$$

$$\theta = \frac{\pi e^2}{k}. \tag{7.33}$$

A conventional method to show that (7.32) is associated with fractional statistics starts with the hypothesis that the wave functions ψ_0 associated with H_θ are functions of $z_{(\alpha)a}$ with a definite symmetry property under exchange. Let us assume for instance that ψ_0 is a symmetric function of $\mathbf{x}_{(\alpha)}$. Now H_θ for a generic θ is not the usual free particle Hamiltonian

$$H_0 = \frac{1}{2m} \sum_\alpha [p_{(\alpha)a}]^2 \qquad (7.34)$$

because of the presence of θ terms. But we can bring H_θ to H_0 by a unitary transformation:

$$H_0 = U_\theta H_\theta U_\theta^{-1},$$

$$U_\theta = \prod_{\alpha<\beta} \exp\left[i\frac{\theta}{\pi}\phi_{\alpha\beta}\right]. \qquad (7.35)$$

In the representation where the Hamiltonian is H_0, the wave functions are

$$\phi_\theta = U_\theta \psi_0. \qquad (7.36)$$

Next let us consider what happens to ψ_θ if we exchange particles α and β by moving $\mathbf{z}_{(\alpha)}$ to $\mathbf{z}_{(\beta)}$ and $\mathbf{z}_{(\beta)}$ to $\mathbf{z}_{(\alpha)}$.

Let us perform this exchange so that $\mathbf{z}_{(\alpha)} - \mathbf{z}_{(\beta)}$ rotates by π and not by $-\pi$. Then $\phi_{\alpha\beta}$ becomes $\phi_{\alpha\beta}(t) + \pi$ whereas this exchange can be so performed that the remaining terms $\sum \phi_{\gamma\delta}$ [with the particle pair $(\gamma\delta)$ distinct from the pair $(\alpha\beta)$] are unaffected. Thus ψ_α transforms under this exchange according to

$$\psi_\theta \to e^{i\theta}\psi_\theta. \qquad (7.37)$$

The phase $e^{i\theta}$ may be said to characterize the particle statistics.

There is of course no reason why we could not have assumed that the wave functions on which H_θ operates are of the type ψ_δ. Under the preceding exchange, ψ_δ transforms according to (7.37) with θ replaced by δ. Had we done so, the statistical parameter for the wave functions on which H_0 operates would have been $\theta + \delta$ instead of θ. In other words, we can with equal legitimacy assume that fractional statistics is due in part to the Chern-Simons term and in part to the properties of the wave functions. By convention, it is assumed in the literature that it arises exclusively from the Chern-Simons term. This amounts to the convention that the wave functions on which H_θ operates obey Bose or Fermi statistics if the particle spins are integral or half odd integral in the absence of the Chern-Simons term. In the latter case, the wave functions contribute an amount π to the statistical parameter. We shall see below that particles of no intrinsic spin in the absence of the Chern-Simons term may acquire such a spin in the presence of this term.

For further discussion of these points, see Sec. 2 of Ref. [24].

7.4. *Spin from the Chern-Simons term*

The rotation group associated with the plane $\mathbb{R}^2$ is SO(2). In quantum mechanics, we have the possibility that wave functions do not carry a true representation of a symmetry group, but only a representation up to a phase.

Often this means that they transform by a representation of the universal covering group of the symmetry group. Now SO(2) is isomorphic to the group $U(1) = \{e^{i\phi}\}$. It has the topology of a circle and is infinitely connected. Its universal covering group therefore covers it infinitely often. This covering group is $\mathbb{R}^1 = \{a\}$, the group law being addition. The covering homomorphism is $a \to e^{i2\pi a}$. The kernel of this homomorphism, or the elements of $\mathbb{R}^1$ which map to the identity, is $\mathbb{Z} = \{n\}$, n being an integer. This group $\mathbb{Z}$ being the inverse image of for instance the 2π rotations, we have the possibility of wave functions ψ which acquire a phase under 2π rotation $R_{2\pi}$:

$$R_{2\pi}\psi = e^{i\theta}\psi . \tag{7.38}$$

As these "rotations" are supposed to form a representation of $\mathbb{Z}$, no power of $e^{i\theta}$ need to be 1 and this phase is not subject to any restriction.

Let R_a denote the operator for the element a of $\mathbb{R}^1$. Its action on ψ compatible with (7.38) is

$$R_a\psi = e^{i\theta a/2\pi}\psi . \tag{7.39}$$

If S is the spin of the particle, we normally write

$$R_a = e^{iaS} . \tag{7.40}$$

Hence:

$$S = \frac{\theta}{2\pi} \tag{7.41}$$

for a particle which transforms according to (7.39).

Since $a \to R_a$ is supposed to be a representation of $\mathbb{R}^1$, there is no constraint on the value of θ [besides reality which comes from requiring unitarity]. Thus we have the possibility of arbitrary fractional spin in 2+1 dimensions. In the ensuing discussion, we shall show that any fractional spin can be induced by an appropriate Chern-Simons term.

In the discussion of particle statistics in the preceding pages, we neglected the self-interaction of the particles in the presence of the Chern-Simons term. It is this term which induces the additional spin on the particles. We now demonstrate this assertion ([30], [31] and references therein).

The field A_μ for a single source α with trajectory $\mathbf{z}_{(\alpha)}(t)$ is

$$A_\mu = -\frac{e}{k}\partial_\mu\phi[\mathbf{x} - \mathbf{z}_{(\alpha)}(t)] .$$

It involves the angle $\phi[\mathbf{x} - \mathbf{z}_{(\alpha)}(t)]$ which does not have a well defined limit as we approach the source. We can describe the ambiguity of this limit by "framing" the source. Thus let us attach a frame to the source, with which a pair of linearly independent vectors is associated. We can choose these vectors to be right-handed and orthonormal with respect to the Euclidean metric. This frame then defines a Cartesian coordinate system with origin at the particle position. The ambiguity in the limiting form of ϕ can be described in terms of the azimuthal angle ψ of the direction of approach to the origin of one such limit. Clearly for a limit characterized by ψ, the limit of $A_\mu dx^\mu$ is

$$A(\psi) = -\frac{e}{k} d\psi . \tag{7.42}$$

The additional term in the Lagrangian induced by self-interaction is thus

$$L_{\text{spin}} = -\frac{e^2}{2k} \frac{d\psi(t)}{dt} \tag{7.43}$$

where a factor $1/2$ mentioned after (7.30) has been included.

Now a rotation of the frame by an angle λ with the particle position as the centre changes ψ to $\psi + \chi$. The momentum p_ψ conjugate to ψ is hence the intrinsic spin S of the particle. Its value follows from (7.43):

$$p_\psi = S = -\frac{e^2}{2k} . \tag{7.44}$$

7.5. *Parity and time reversal*

Parity P and time reversal T are not symmetries of a system of identical particles which obey fractional statistics unless the statistical parameter is 0 or π (mod 2π). These exceptional cases correspond to particles obeying Bose or Fermi statistics. We now briefly explain how this violation of P and T comes about.

The abelian Chern-Simons Lagrangian with identical sources defines a theory which is periodic in the statistical parameter θ with period 2π. This is because, as we have seen, the effect of the Chern-Simons term can be regarded as altering the statistical properties of the states without affecting the Hamiltonian, the states getting multiplied by the phase $e^{i\theta}$ under an interchange σ_i of two sources. As this phase is periodic in θ with period 2π, the periodicity in θ of the theory follows as well.

It is now easy to see that the Chern-Simons Lagrangian with sources violates P and T unless $\theta = 0$ or $\theta = \pi$ (mod 2π), for the Chern-Simons term reverses in sign under either of these operations, in the case of P because of Levi-Civita symbol and in the case of T because it is linear in time derivative or in A_0. [Note here that, in 2+1 dimensions, parity reverses the sign of only *one* of the coordinates, say x_1 or x_2, and not both since the matrix of the transformation $x_1 \to x_1$, $x_2 \to -x_2$ has -1 as determinant and reverses spatial orientation; the transformation $x_1 \to -x_1$, $x_2 \to -x_2$ has neither of these properties, being a π rotation].

It is equally easy to see P and T violation for $\theta \neq 0, \pi$ (mod 2π) in terms of states. Suppose that σ_i corresponds to exchanging particles i and $i + 1$ by moving them in an anti-clockwise direction. Under P, it is then transformed to the exchange $i \leftrightarrow i + 1$ where the particles are moved in a clockwise direction. Thus $P\sigma_i P^{-1} = \sigma_i^{-1}$ so that P is violated if $e^{i\theta} \neq e^{-i\theta}$ or $\theta \neq 0, \pi$ (mod 2π). As regards T, it is antiunitary and maps a state with an eigenvalue $e^{i\theta}$ for σ_i to one with eigenvalue $e^{-i\theta}$. T violation for $\theta \neq 0, \pi$ (mod 2π) thus also follows.

The periodicity of the physical theory in θ has been claimed here only for the Lagrangian density (7.1). It is not true for instance in the presence of a term constant $\times F_{\mu\nu} F^{\mu\nu}$ as it is readily seen from the expression for the mass of $F_{\mu\nu}$ from (7.16). Indeed for a general Lagrangian with a Chern-Simons term and sources, it is not possible to fully account for the effects of the Chern-Simons term by changing the statistics of the source states. These remarks should be kept in mind when discussing such a Lagrangian.

7.6. $\mathbb{C}P^1$ *models as gauge theories*

$\mathbb{C}P^1$ models are alternatively known as SO(3) nonlinear σ-models. They are field theories involving fields which take values on a two-sphere S^2 ($\equiv \mathbb{C}P^1$). Such models emerge naturally in the continuous limit of Heisenberg spin systems on lattices and are therefore of interest in the context of the present review.

In this section, we shall briefly indicate how to formulate any such model as a gauge theory [37, 53, 60]. This formulation is useful in the construction of its Chern-Simons term. It can be generalized to all models wherein the fields take values in a coset space G/H, G and H ($\subset G$) being simple or semisimple Lie groups. For a discussion of these generalizations see [37, 53, 160]. In this section, we shall also briefly discuss $\mathbb{C}P^1$ solitons and the significance of the Chern-Simons term for the physical properties of this soliton.

As remarked above, the physical fields of the $\mathbb{C}P^1$ models are maps $\mathbf{n}$ from space-time [with coordinates $\mathbf{x}, t$] to S^2:

$$\mathbf{n}(\mathbf{x}, t) \cdot \mathbf{n}(\mathbf{x}, t) = 1 . \qquad (7.45)$$

In the gauge theory formulation of this model, one introduces a pair $z = (z_1, z_2)$ of complex fields on space-time which obey the constraint

$$z^\dagger(\mathbf{x}, t) z(\mathbf{x}, t) \equiv \sum_{\alpha=1}^{2} z_\alpha^*(\mathbf{x}, t) z_\alpha(\mathbf{x}, t) = 1 . \qquad (7.46)$$

This may be written as

$$\sum_{\alpha, a=1}^{2} [\xi_\alpha^a(\mathbf{x}, t)]^2 = 1 \qquad (7.47)$$

by separating the real and the imaginary parts of z_α [$z_\alpha = \xi_\alpha{}^1 + \xi_\alpha{}^2$] whereupon it is clearly seen to describe the three-sphere S^3. Hence, z takes values on S^3. The physical σ-model field **n** can be recovered by using the formula

$$n_\alpha = z^\dagger \sigma_\alpha z , \qquad \sigma_\alpha = \text{Pauli matrices} . \tag{7.48}$$

The constraint (7.45) on **n** is then automatically fulfilled because of the property (7.46) of z.

The vector $z^\dagger(\mathbf{x}, t)\boldsymbol{\sigma} z(\mathbf{x}, t)$ is invariant under the U(1) gauge transformation

$$z(\mathbf{x}, t) \to e^{i\theta(\mathbf{x}, t)} z(\mathbf{x}, t) . \tag{7.49}$$

Now when we identify all points of S^3 related by such an action of U(1), the result is S^2. In mathematics, the manifold S^3 is said to be a U(1) principal bundle (the Hopf bundle) over S^2 with structure group U(1) to express essentially this relation between S^3 and S^2 [167]. We remark that **n** may be regarded as a point of this S^2.

A standard Lagrangian density for **n** (in the absence of the Chern-Simons term) is

$$\mathcal{L}_0 = \text{constant} \times [\partial_\mu \mathbf{n}(\mathbf{x}, t)]^2 \tag{7.50}$$

where the constant has the dimension of energy in 2+1 dimensions.

$\mathcal{L}_0$ can be regarded as a function of $\mathbf{z}(\mathbf{x}, t)$ by substituting the definition (7.48). We shall now manufacture a U(1) gauge field A_μ out of z. It is just

$$A_\mu = -iz^\dagger \partial_\mu z . \tag{7.51}$$

It is easy to see that A_μ transforms as a potential

$$A_\mu \to A_\mu + \partial_\mu \theta \tag{7.52}$$

under the gauge transformation (7.49) because of the constraint (7.46).

The Chern-Simons term for the $\mathbb{C}P^1$ model is constructed from A_μ. It reads

$$\mathcal{L}_{\text{CS}} = \frac{k}{4\pi} \epsilon^{\mu\nu\lambda} A_\mu \partial_\nu A_\lambda \tag{7.53}$$

where A_μ is to be thought of as a function of z while performing, for example, variational calculations.

An important property of this term is that it has no effect on the field equations. This may be shown by first noting that the general variation δz of z fulfills

$$\delta z^\dagger(\mathbf{x}, t) z(\mathbf{x}, t) + z^\dagger(\mathbf{x}, t) \delta z(\mathbf{x}, t) = 0 , \tag{7.54}$$

in view of (7.46). Using this equation and its derivative, it can be shown after some calculations that the general variation of $\mathcal{L}_{\text{CS}}$ is a total divergence, and hence does not affect the field equations. Another way to show this result is to note that AdA is a three-form on the three manifold S^3 and hence closed. Its variation under an infinitesimal variation of z will therefore be by an exact form.

The $\mathbb{C}P^1$ model has solitons which are in a certain sense the 2+1 dimensional analogues of the Skyrme solitons [28]. We can see the possibility of these solitons by examining the topology of the configuration space of this model. Its equations of motion involve only the field $\mathbf{n}$. The physical field is thus to be regarded as $\mathbf{n}$ and not z. Indeed it is $\mathbf{n}$ and not z which is invariant under the gauge transformation (7.49). Thus the configuration space $\mathcal{Q}$ of this model consists of maps of $\mathbb{R}^2$ with coordinates $\mathbf{x}$ into S^2. Finiteness of energy requires the existence of the integral

$$\int [\partial_i \mathbf{n}(\mathbf{x}, t)]^2 \qquad (7.55)$$

which is guaranteed if

$$\mathbf{n}(\mathbf{x}, t) \to \text{constant} \qquad \text{as } r = |\mathbf{x}| \to \infty \ . \qquad (7.56)$$

Standard arguments show that there is no loss of generality in choosing this constant to be $(0, 0, 1)$:

$$\mathbf{n}(\mathbf{x}, t) \to (0, 0, 1) \qquad \text{as } r \to \infty \ . \qquad (7.57)$$

As the field $\mathbf{n}$ takes a constant value as $r \to \infty$, we can also identify all "points at infinity" $\mathbb{R}^2$ and regard it as S^2. Thus the configuration space may be thought of as consisting of maps of S^2 to S^2:

$$\mathcal{Q} = \text{Maps}(S^2, S^2) \ . \qquad (7.58)$$

This space $\mathcal{Q}$ is not connected. It consists of an infinite number of disconnected pieces $\mathcal{Q}_n$ labelled by a winding number N:

$$\mathcal{Q} = \bigcup_n \mathcal{Q}_n \ , \qquad N \in \mathbb{Z} \ . \qquad (7.59)$$

The expression for N in terms of the field variables is

$$N = \frac{1}{2\pi} \int F_{12} d^2x = \frac{1}{8\pi} \int \epsilon^{ab} \epsilon_{ijk} n^i \partial_a n^j \partial_b n^k \ , \qquad (7.60)$$

$$F_{12} = \partial_1 A_2 - \partial_2 A_1 \ , \qquad (7.61)$$

$$\epsilon^{12} = -\epsilon^{21} = 1 \ . \qquad (7.62)$$

The function N, being integer valued, is a constant of motion. It is thus conserved for topological reasons.

The topological sector with $N = 0$ is the vacuum sector. The sectors with $N = \pm 1$ may be thought of as associated with an elementary soliton and antisoliton since sectors with $|N| \geq 2$ can be obtained by composing these sectors in an appropriate sense.

It is important to know if there are static solitons in this model. If they exist, then they are candidate solutions which on quantization may admit an interpretation as particle-like excitations. There are in fact such static solutions for any N.

The action associated with the Lagrangian density (7.50) is invariant under the scale transformation

$$z(\mathbf{x}, t) \to z(\lambda \mathbf{x}, t), \qquad \lambda \in \mathbb{R}^1 - \{0\}, \tag{7.63}$$

so that the scale transform of a static solution is still a static solution. This has the consequence that the Lagrange function (7.50) does not classically yield a specific value for soliton size. This is at first sight a problem for the particle interpretation of the solitons. It can however be resolved by augmenting the Lagrange function by scale breaking terms. Alternatively, as in the Skyrme model [28, 35, 36, 113], one can introduce a collective coordinate for the scale transformation. It may be that the mean value of this coordinate in the quantum ground state is fixed thereby removing the classical uncertainty regarding the soliton size.

The spin and statistics of the solitons are affected by the Chern-Simons term, the induced spin and the statistical parameter being proportional to k. We shall not show these results here but refer to Wilczek and Zee [185, 189] where a detailed treatment may be found.

We conclude this section by outlining an alternative, but entirely equivalent form of the $\mathbb{C}P^1$ Chern-Simons term. There are advantages in this formulation as it admits an easy generalization from $S^2 = SU(2)/U(1)$ to general coset spaces G/H [37, 160]. For the case at hand, where $G = SU(2)$ and $H = U(1)$, let us regard G concretely as 2×2 unitary matrices of unit determinant and $U(1)$ as the subgroup with generator σ_3. The Lagrangian is written using fields g with values in $SU(2)$. Associated with g, we also introduce the field $\mathbf{n}$ via the definition

$$n_\alpha \sigma_\alpha = g \sigma_3 g^\dagger \tag{7.64}$$

so that

$$n_\alpha(\mathbf{x}, t) = \frac{1}{2} \text{Tr}[\sigma_\alpha g(\mathbf{x}, t) \sigma_3 g(\mathbf{x}, t)^\dagger]. \tag{7.65}$$

A simple calculation shows that $n_\alpha(\mathbf{x}, t) n_\alpha(\mathbf{x}, t) = 1$. It is thus consistent to identify this $\mathbf{n}$ with the unit vector field of the nonlinear σ-model.

It is important to note that $\mathbf{n}$ is invariant under the $U(1)$ gauge transformation

$$g(\mathbf{x}, t) \to g(\mathbf{x}, t) e^{i \sigma_3 \theta(\mathbf{x}, t)}, \tag{7.66}$$

which is the analogue of the transformation (7.49) in the present formalism.

Now from g, we can construct a U(1) gauge potential A_μ by setting

$$A_\mu(\mathbf{x}, t) = -\frac{1}{2} i \, \text{Tr}[\sigma_3 g(\mathbf{x}, t)^{-1} \partial_\mu g(\mathbf{x}, t)] \,. \tag{7.67}$$

It transforms under (7.66) as

$$A_\mu \to A_\mu + \partial_\mu \theta \,. \tag{7.68}$$

This potential is the analogue of (7.51).

The Chern-Simons term can now be written as

$$\mathcal{L}_{\text{CS}} = \frac{k}{4\pi} \epsilon^{\mu\nu\lambda} A_\mu F_{\nu\lambda} \,, \tag{7.69}$$

$$F_{\nu\lambda} = \partial_\nu A_\lambda - \partial_\lambda A_\nu \,. \tag{7.70}$$

The kinetic energy term of $\mathbf{n}$ can also be written using g. It is a simple exercise [37, 160] to show that

$$-\frac{1}{2} [\partial_\mu \mathbf{n}]^2 = \text{Tr}[g D_\mu g^\dagger] \,, \tag{7.71}$$

$$D_\mu g^\dagger := \partial_\mu g^\dagger + i A_\mu \sigma_3 g^\dagger \,. \tag{7.72}$$

These two formalisms using z and g are actually the same. To see this write g as

$$g = \begin{bmatrix} z_1 & -z_2^* \\ z_2 & z_1^* \end{bmatrix}, \tag{7.73}$$

where the field z satisfies (7.46). Now

$$\text{Tr}[g(\mathbf{x}, t)^{-1} \partial_\mu g(\mathbf{x}, t)] = 0 \,, \tag{7.74}$$

since $g^{-1} \partial_\mu g$ takes values in the SU(2) Lie algebra which is spanned by traceless matrices. Therefore

$$\text{Tr}[\sigma_3 g^{-1} \partial_\mu g] = \text{Tr}[(\sigma_3 + 1) g^\dagger \partial_\mu g]$$

$$= \text{Tr}\left[\begin{bmatrix} 2 & 0 \\ 0 & 0 \end{bmatrix} g^\dagger \partial_\mu g\right]$$

$$= 2 g_{\alpha 1}^* \partial_\mu g_{\alpha 1} = 2 z_\alpha^* \partial_\mu z_\alpha \,, \tag{7.75}$$

so that the potentials and hence the Chern-Simons term in the two formalisms are the same.

8. The Non-Abelian Chern-Simons Term

The non-abelian Chern-Simons terms can be constructed following an idea explained earlier in the abelian case. Let G be a compact simple Lie group which we think of concretely as a group of matrices [the construction generalizes trivially to semisimple Lie groups or to Lie groups with the Lie algebras which admit an invariant bilinear form]. The non-abelian connection A_μ is now valued in the Lie algebra $\underline{G}$ of G. If $T(\alpha)$ $\{\alpha = 1, \ldots, [G]; [G] =$ dimension of $G\}$ is a hermitian basis for $\underline{G}$, we can write $A_\mu = A^\alpha_\mu T(\alpha)$. In our convention in what follows, $A^\alpha_\mu(\mathbf{x}, t)$ are purely imaginary. The curvature tensor

$$F_{\mu\nu} = \partial_\mu A_\nu - \partial_\nu A_\mu + [A_\mu, A_\nu] \tag{8.1}$$

associated with A_μ fulfills the Bianchi identity

$$D_\mu F_{\nu\lambda} = \partial_\mu F_{\nu\lambda} + [A_\mu, F_{\nu\lambda}] = 0 . \tag{8.2}$$

If we introduce the forms

$$A = A_\mu dx^\mu , \tag{8.3}$$

$$F = F_{\mu\nu} dx^\mu \wedge dx^\nu , \tag{8.4}$$

this equation can be written as

$$DF = 0 , \qquad DF := dF + A \wedge F - F \wedge A . \tag{8.5}$$

These equations are valid in all space-time dimensions. We will not restrict space-time dimensions till (8.14).

Now consider $d \operatorname{Tr}(F \wedge F)$. It equals $\operatorname{Tr}(dF \wedge F) + \operatorname{Tr}(F \wedge dF)$. Using (8.5) we have

$$d \operatorname{Tr}(F \wedge F) = \operatorname{Tr}(DF \wedge F) + \operatorname{Tr}(F \wedge DF) = 0 . \tag{8.6}$$

In other words, $\operatorname{Tr}(F \wedge F)$ is closed and can be written as

$$\operatorname{Tr}(F \wedge F) = d\left[\frac{\pi}{k} \Omega^{(3)}\right] . \tag{8.7}$$

A particular solution $\omega^{(3)}$ for $\Omega^{(3)}$ is readily found, and it is the non-abelian Chern-Simons term (written as a three-form):

$$\omega^{(3)} = \frac{k}{4\pi} \operatorname{Tr}\left(A \wedge dA + \frac{2}{3} A \wedge A \wedge A\right) . \tag{8.8}$$

The general solution for $\Omega^{(3)}$ can be written as

$$\Omega^{(3)} = \omega^{(3)} + d\phi \tag{8.9}$$

where ϕ is a two-form.

Remembering that, under the gauge transformation

$$A \to g^{-1}Ag + g^{-1}dg, \tag{8.10}$$

of A, F transforms as

$$F \to g^{-1}Fg, \tag{8.11}$$

we see that $\text{Tr}(F \wedge F)$ is gauge invariant. As a consequence, $\omega^{(3)}$ can change only by a closed form χ. An explicit calculation verifies this assertion to be correct. The expression for χ is

$$\chi = -\frac{k}{12\pi} \text{Tr}(g^{-1}dg \wedge g^{-1}dg \wedge g^{-1}dg) + \frac{k}{4\pi} d\,\text{Tr}(A \wedge dgg^{-1}). \tag{8.12}$$

It is a simple matter to verify that $d\chi = 0$.

The standard non-abelian Chern-Simons Lagrangian function $\mathcal{L}_{\text{CS}}$ follows from $\omega^{(3)}$ by writing $A = A_\mu(x)dx^\mu$ and considering 2+1 dimensional space-times. Then

$$\omega^{(3)} = \mathcal{L}_{\text{CS}} d^3x \tag{8.13}$$

$$\mathcal{L}_{\text{CS}} = \frac{k}{4\pi} \epsilon^{\mu\nu\lambda} \text{Tr}\left(A_\mu \partial_\nu A_\lambda + \frac{2}{3} A_\mu A_\nu A_\lambda\right). \tag{8.14}$$

The last term here is zero in the abelian case.

The response $\omega^{(3)} \to \omega^{(3)} + d\chi$ of $\omega^{(3)}$ under a gauge transformation (8.10) shows that

$$\mathcal{L}_{\text{CS}} \to \mathcal{L}_{\text{CS}} + \frac{k}{4\pi} \partial_\mu [\epsilon^{\mu\nu\lambda} \text{Tr}(A_\nu \partial_\lambda g g^{-1})]$$

$$- \frac{k}{12\pi} \epsilon^{\mu\nu\lambda} \text{Tr}[(g^{-1}\partial_\mu g)(g^{-1}\partial_\nu g)(g^{-1}\partial_\lambda g)]. \tag{8.15}$$

For infinitesimal g [$g \simeq 1 + i\epsilon^\alpha T(\alpha)$ where $T(\alpha)$ are the Lie algebra generators], the last term is also a total divergence (as it is readily verified). The equations of motion from $\mathcal{L}_{\text{CS}}$ are thus gauge invariant. Explicitly, they are

$$F_{\mu\nu} = 0. \tag{8.16}$$

Again the last term here is zero in the abelian case. It has the remarkable property of constraining the value of k in the non-abelian problem. To see how this comes about, let us specialize first to $G = SU(2)$ where we regard $SU(2)$ as 2×2 unitary unimodular matrices. We also impose the usual boundary condition

$$A_\mu(\mathbf{x}, t) \to 0 \text{ at space-time infinity}, \qquad (8.17)$$

A gauge transformation g of A going to 1 at ∞ respects this condition. Consider the group $\mathcal{G}^\infty$ of such transformations:

$$\mathcal{G}^\infty = \{g \,|\, g(\mathbf{x}, t) \in SU(2), g(\mathbf{x}, t) \to 1 \text{ at } \infty\}. \qquad (8.18)$$

Because of the boundary condition, it can be regarded as the group of maps of the three sphere S^3 to $SU(2)$. As it is well known, for example from Skyrmion physics [28], these maps fall into an infinite number of disjoint pieces $\mathcal{G}^\infty_n$ classified by an integer valued winding number N:

$$\mathcal{G}^\infty = \amalg \mathcal{G}^\infty_n, \qquad (8.19)$$

$$N = \frac{1}{24\pi^2} \int \epsilon^{\mu\nu\lambda} \mathrm{Tr}[(g^{-1}\partial_\mu g)(g^{-1}\partial_\nu g)(g^{-1}\partial_\lambda g)] d^3x, \qquad (8.20)$$

$$N \in \mathbb{Z}. \qquad (8.21)$$

Therefore, for a gauge transformation with $N \neq 0$, the action

$$S_{CS} = \int d^3x \mathcal{L}_{CS} \qquad (8.22)$$

changes according to

$$S_{CS} \to S_{CS} - 2\pi k N. \qquad (8.23)$$

In other words, S_{CS} is not gauge invariant.

In the functional integral form of quantum theory, it is not the action itself, but rather $\exp[iS_{CS}]$ that occurs [we set $\hbar = 1$]. We can therefore restore the gauge invariance of quantum theory by requiring that

$$e^{-i2\pi k n} = 1 \qquad (8.24)$$

or that

$$k \in \mathbb{Z}, \qquad (8.25)$$

we thus see that gauge invariance forces the coefficient k in the Chern-Simons term to be quantized if $G = \mathrm{SU}(2)$.

There is a similar quantization condition for any simple Lie group G. For $\mathrm{SU}(N)$, with G realized by its $N \times N$ unitary irreducible representation (UIR), and with the trace in (8.14) referring to this UIR, the condition on k is still (8.25). In general it is not (8.25), but can be worked out in each specific case.

There is an interpretation of the condition (8.25) in the canonical approach. Briefly stated, it is the following. In the canonical approach, the gauge "invariance" of the Lagrangian leads to the Gauss law constraint, and we require that the quantum states are annihilated by this constraint. The Lie algebra generated by these constraints is the Lie algebra of the gauge group G^∞ obtained by freezing time in $\mathscr{G}^\infty$:

$$G^\infty = \{g \mid g(\mathbf{x}) \in G,\ g(\mathbf{x}) \to 1 \text{ as } |\mathbf{x}| \to \infty\} \ . \tag{8.26}$$

There is however no certainty that the group generated by the Gauss law is precisely G^∞. This is due to the following circumstance. Because of the boundary condition, G^∞ can be regarded as the group of maps of S^2 to G. This group is infinitely connected for a simple Lie group G:

$$\pi_1(G^\infty) = \pi_3(G) = \mathbb{Z} \ . \tag{8.27}$$

[If $G = \mathrm{U}(1)$, $\pi_1(G^\infty) = \{0\}$ and this is why there is no constraint on k in the abelian case]. Hence, the universal covering group $\bar{G}^\infty$ of G^∞ is a group which covers it infinitely often. The centre of $\bar{G}^\infty$ is $\mathbb{Z}$ and

$$G^\infty = \frac{\bar{G}^\infty}{\mathbb{Z}} \ . \tag{8.28}$$

The group G^∞ and $\bar{G}^\infty$ have precisely the same Lie algebras. It may happen that the Gauss law does not generate G^∞, but instead generates $\bar{G}^\infty/\mathbb{Z}'$ where $\mathbb{Z}'$ is a proper subgroup of $\mathbb{Z}$. In either of these cases, the centre of $\bar{G}^\infty$ is represented by a nontrivial phase in quantum theory whereas the centre of $\bar{G}^\infty$ must be represented by the unit operator on all states ("physical" states) annihilated by the Gauss law. The theory thus becomes inconsistent when the Gauss law generates $\bar{G}^\infty$ or $\bar{G}^\infty/\mathbb{Z}'$.

Now it may be shown that the centre of $\bar{G}^\infty$ on quantum states consists precisely of the factors $\exp[-i2\pi k N]$, picked up by $\exp[iS_{\mathrm{CS}}]$, under the gauge transformations g. These factors must therefore be 1 for reasons of consistency, which requires in turn that k must be quantized in suitable units as discussed earlier.

8.1. *Statistics and Spin of Identical Sources*

We will now investigate the statistics and spin of N identical sources in interaction with a non-abelian gauge field described by a Chern-Simons term. As we have seen, in the corresponding abelian problem, the abelian sources (charged particles) for fixed charges will as a rule obey a generalized statistics determined by the strength θ of the Chern-Simons term. Bose and Fermi statistics are recovered only for certain discrete values of the parameter θ. Our investigation here will as before follow the methodology of earlier work and in particular that of Arovas et al. [27, 30, 31]. We shall thus first find all solutions of the field equations with sources. The field is then eliminated essentially by substituting a solution in the Lagrangian thereby obtaining an effective Lagrangian L_{eff} for the sources. The statistics and spin of the sources are then inferred by quantization of L_{eff}. The work is due to Refs. [30, 31].

We will concentrate on the gauge group SU(2) first for reasons of simplicity, and second because it appears to be the group relevant for the strongly coupled half-filled Hubbard model. We will also be brief on certain details. A detailed treatment [covering also groups other than SU(2)] may be found in Ref. [31].

The sources for the SU(2) gauge fields are assumed to be point particles transformed by the $(2j + 1)$ dimensional unitary irreducible representation (UIR) $\Gamma^{(j)}$ of SU(2) $[j \in \{0, 1/2, 1, \ldots\}]$. The Lagrangian for one such particle [call it α] in the absence of gauge fields is known to be [167]:

$$L_{(\alpha)} = \frac{1}{2} m \dot{x}_{(\alpha)}(t)^2 + ij \, \text{Tr}[\tau_3 s_{(\alpha)}(t)^{-1} \dot{s}_{(\alpha)}(t)] \, , \tag{8.29}$$

$$x_{(\alpha)}(t) \in \mathbb{R}^2 \, , \qquad s_{(\alpha)}(t) \in \text{SU}(2) \, .$$

Here m is the particle mass, τ_j are Pauli matrices and the group SU(2) is concretely thought of as the group of 2×2 unitary unimodular matrices. The variation of $x_{(\alpha)}(t)$ gives free motion in $x_{(\alpha)}(t)$ $[\ddot{x}_{(\alpha)}(t) = 0]$ while the variation $\delta s_{(\alpha)}(t) = -i[\epsilon_j(t)\tau_j/2]s_{(\alpha)}(t)$ shows that

$$\dot{I}_{(\alpha)}(t) = 0 \, , \tag{8.30}$$

$$I_{(\alpha)}(t) := I_{(\alpha)j}(t)\tau_j = js_{(\alpha)}(t)\tau_3 s_{(\alpha)}(t)^{-1} \, .$$

On quantization, $I_{(\alpha)j}$ become associated with quantum operators $\hat{I}_{(\alpha)j}$ obeying the commutation relations $[\hat{I}_{(\alpha)i}, \hat{I}_{(\alpha)j}] = i\epsilon_{ijk}\hat{I}_{(\alpha)k}$ and generating the UIR $\Gamma^{(j)}$. The Lagrangian L_{source} for N such particles is obtained by adding $L_{(\alpha)}$.

In the presence of a gauge field A, L_{source} must be modified to $L'_{\text{source}} = \sum L'_{(\alpha)}$ where

$$L'_{(\alpha)} = \frac{1}{2} m \dot{z}_{(\alpha)}{}^a(t) \dot{z}_{(\alpha)}{}^a(t) + ij \, \text{Tr}[\tau_3 s_{(\alpha)}(t)^{-1} D_t s_{(\alpha)}(t)] \, ,$$

$$D_t s_{(\alpha)}(t) = \dot{s}_{(\alpha)}(t) + \dot{z}_{(\alpha)}{}^\mu(t) A_\mu[z_{(\alpha)}(t)] s_{(\alpha)}(t) \, ,$$

$$z_{(\alpha)}{}^a(t) = x_{(\alpha)}{}^a(t) \, , \qquad a = 1, 2 \, ; \qquad z_{(\alpha)}{}^0(t) = t \, . \tag{8.31}$$

[Here and in what follows, Roman and Greek space-time indices take values 1, 2 and 0, 1, 2 respectively.] We may furthermore augment this Lagrangian by the Chern-Simons Lagrangian L_{CS} for A:

$$L_{CS} = \int d^2x \mathcal{L}_{CS} \tag{8.32}$$

$$\mathcal{L}_{CS} = \frac{k}{4\pi} \epsilon^{\mu\nu\lambda} \text{Tr}\left[A_\mu \partial_\nu A_\lambda + \frac{2}{3} A_\mu A_\nu A_\lambda\right], \quad k \in \mathbb{Z}.$$

The variations of $s_{(\alpha)}$, $x_{(\alpha)}$ and A now lead to the analogues of Wong's Eqs. [167]:

$$\dot{I}_{(\alpha)}(t) + \dot{z}_{(\alpha)}{}^\mu(t)[A_\mu(z_{(\alpha)}(t)), I_{(\alpha)}(t)] := D_t I_{(\alpha)}(t) = 0, \tag{8.33}$$

$$m\ddot{x}_{(\alpha)}{}^a(t) = i \, \text{Tr}[I_{(\alpha)}(t) F_\lambda{}^a(z_{(\alpha)}(t))]\dot{z}_{(\alpha)}{}^\lambda(t), \tag{8.34}$$

$$\frac{k}{2\pi} F_{ab}(x) = -i\epsilon_{ab} \sum_\alpha \delta^2[x - x_{(\alpha)}(t)]I_{(\alpha)}(t), \tag{8.35}$$

$$\frac{k}{2\pi} F_{a0}(x) = i\epsilon_{ab} \sum_\alpha \delta^2[x - x_{(\alpha)}(t)]\dot{z}_{(\alpha)}{}^b(t)I_{(\alpha)}(t), \tag{8.36}$$

$$F_{\mu\nu} = \partial_\mu A_\nu - \partial_\nu A_\mu + [A_\mu, A_\nu],$$

$$\epsilon_{ab} = -\epsilon_{ba}, \quad \epsilon_{12} = +1. \tag{8.37}$$

A particular consequence of these equations is that, classically, the motion in the coordinates $x_{(\alpha)}$ is free as seen by substituting Eqs. (8.35)–(8.36) in (8.34). [Quantum mechanically, however, the sources will in general scatter [13, 59].] Note that (8.33) follows from (8.35)–(8.36) and the Bianchi identity.

Let us now describe all solutions of (8.35)–(8.36) and hence (8.33) [modulo certain technical quantifications] in a particular gauge. In describing the solution, we will assume that no two sources have the same first coordinate [$x_{(\alpha)1} - x_{(\beta)1} \neq 0$ if $\alpha \neq \beta$]. This assumption will be relaxed later. Let $\Delta_{(\alpha)}$ be a thin strip from $x_{(\alpha)}(t)$ going along the negative 2 direction. The tails $\Delta_{(\alpha)}$ are taken to be sufficiently thin so as not to overlap [$\Delta_{(\alpha)} \cap \Delta_{(\beta)} = \emptyset$ if $\alpha \neq \beta$]. Let $\psi_{(\alpha)}$ be an angle-like function which (at time t) is constant in $\mathbb{R}^2 \setminus \Delta_{(\alpha)}$ and increases by 2π as $\Delta_{(\alpha)}$ is crossed from left to right. The general solution up to a gauge transformation can be constructed as follows. Set

$$I_{(1)} = K_{(1)} = j\tau_3, \tag{8.38}$$

$$I_{(\alpha)} = K_{(\alpha)} = ju_{(\alpha)}\tau_3 u_{(\alpha)}^{-1}, \quad \alpha \geq 2,$$

$u_{(\alpha)}$ being time independent SU(2) matrices. Then [with $k \neq 0$]:

$$A = A_\mu dx^\mu = \begin{cases} 0 & \text{in } \mathbb{R}^2 \setminus \bigcup_\alpha \Delta_{(\alpha)} \\ -\dfrac{i}{k} K_{(\alpha)} \dfrac{\partial \psi_{(\alpha)}}{\partial x^\mu} dx^\mu & \text{in } \Delta_{(\alpha)} \end{cases} \quad (8.39)$$

The gauge fixing leading to (8.39) is not complete, since for instance a rigid rotation around the third axis does not change $I_{(1)}$ and leads to a gauge equivalent solution. More generally, any two solutions are gauge equivalent [modulo certain technical qualifications] if the associated holonomies are the same conjugancy class. There are no further gauge equivalences among the solutions.

Since

$$\int_{C_{(\alpha)}} \frac{\partial \psi_{(\alpha)}}{\partial x^a} dx^a = 2\pi \quad (8.40)$$

for a counterclockwise loop $C_{(\alpha)}$ enclosing only the source α, we have, by Stokes' theorem:

$$(\partial_a \partial_b - \partial_b \partial_a) \psi_{(\alpha)}(x) = \epsilon_{ab} 2\pi \delta^2[x - x_{(\alpha)}(t)] . \quad (8.41)$$

Using this identity and the fact that $\psi_{(\alpha)}$ can be regarded as a function of $x^a - x_{(\alpha)}{}^a(t)$, it is easy to show that A solves the field equations in the gauge where $I_{(\alpha)} = K_{(\alpha)}$.

Although the solution looks abelian, the holonomics $W_{(\alpha)}$ associated with it need not be so since

$$W_{(\alpha)} = P \exp\left[-\int_{C_{(\alpha)}} A\right] = \exp\left[+\frac{2\pi i}{k} K_{(\alpha)}\right] . \quad (8.42)$$

Clearly different $W_{(\alpha)}$ need not commute for suitable choices of $K_{(\alpha)}$ and k.

In the abelian problem, the solution of (8.35)–(8.36) for given source trajectories is unique up to a gauge transformation. This, as we have seen, is not so in the non-abelian problem. The existence of gauge inequivalent solutions has unusual consequences for source statistics as we shall see below.

Let us for a while ignore self-interactions. The potential seen by particle α due to the remaining sources in an arbitrary gauge is

$$g(z_{(\alpha)}) A_{(\alpha)}(z_{(\alpha)}) g(z_{(\alpha)})^{-1} + g(z_{(\alpha)}) dg(z_{(\alpha)})^{-1} , \quad (8.43)$$

$$A_{(\alpha)}(x) := A_{(\alpha)\mu}(x) dx^\mu = -\frac{i}{k} \sum_{\beta \neq \alpha} K_{(\beta)} \frac{\partial \psi_{(\beta)}(x)}{\partial x^\mu} dx^\mu ,$$

g being a general gauge transformation. Substitution of (8.43) in (8.31) gives the effective Lagrangian L_{eff} in the absence of self-interactions:

$$L_{\text{eff}} = \sum_\alpha L_{(\alpha)\text{eff}} , \quad (8.44)$$

$$L_{(\alpha)\,\text{eff}} = \frac{1}{2} m \dot{z}_{(\alpha)}{}^a \dot{z}_{(\alpha)}{}^a + \frac{1}{2} ij \, \text{Tr}[\tau_3 S_{(\alpha)}{}^{-1} D_t S_{(\alpha)}] \;,$$

$$S_{(\alpha)} = g(z_{(\alpha)})^{-1} s_{(\alpha)} \;,$$

$$D_t S_{(\alpha)} = \dot{S}_{(\alpha)} + A_{(\alpha)\mu}(z_{(\alpha)}) \frac{dz_{(\alpha)}{}^\mu}{dt} S_{(\alpha)} \;.$$

The factor $1/2$ in the second term of L_{eff} arises for reasons indicated after (7.30).

L_{eff} involves N fields $A_{(\alpha)t} = A_{(\alpha)\mu}(z_{(\alpha)}) dz_{(\alpha)\mu}/dt$. The redefinition $S_{(\alpha)} = T_{(\alpha)\mu}{}^{-1} S'_{(\alpha)}$ transforms them into $A'_{(\alpha)t} = T_{(\alpha)}[d/dt + A_{(\alpha)t}] T_{(\alpha)}{}^{-1}$.

Hence the N-tuple $\{A_{(\alpha)t}\} = \{A_{(1)t}, A_{(2)t}, \ldots, A_{(n)t}\}$ of these fields related by such transformations lead to equivalent dynamics and can be regarded as gauge equivalent for L_{eff}.

We now verify that (8.44) gives reasonable equations of motion.

The solution (8.39) is valid if no two tails $\Delta_{(\alpha)}$, $\Delta_{(\beta)}$ ($\alpha \neq \beta$) cross. Thus $L_{(\alpha)\,\text{eff}}$ is correct if source α never crosses the common tail of two sources β and γ with the same first two coordinates, that is if $\Delta_{(\alpha)} \cap \Delta_{(\beta)} \cap \Delta_{(\gamma)} = \emptyset$ (α, β, γ unequal). In other words, $L_{(\alpha)\,\text{eff}}$ is correct [in the limit of infinitely thin $\Delta_{(\alpha)}$'s] if no three particles have the same first coordinate.

Variation of $S_{(\alpha)}$ leads to

$$\dot{I}_{(\alpha)}(t) + \frac{1}{2} \dot{z}_{(\alpha)\mu}[A_{(\alpha)\mu}(z_{(\alpha)}), I_{(\alpha)}(t)] := D_t I_{(\alpha)} = 0 \tag{8.45}$$

where here $I_{(\alpha)} = j S_{(\alpha)} \tau_3 S_{(\alpha)}{}^{-1}$. Now for a given position of α for which L_{eff} is valid, at most one term in the sum in $A_{(\alpha)}$ contributes since $\Delta_{(\alpha)} \cap \Delta_{(\beta)} \cap \Delta_{(\gamma)} = \emptyset$.

Therefore either

$$\dot{I}_{(\alpha)}(t) = 0 \tag{8.46a}$$

or

$$\dot{I}_{(\alpha)}(t) - \frac{1}{2} \frac{i}{k} \frac{d\psi_{(\beta)}(z_{(\alpha)})}{dt} [K_{(\beta)}, I_{(\alpha)}(t)] = 0 \tag{8.46b}$$

for some fixed $\beta \neq \alpha$ [(8.44) has been written using the fact that $\psi_{(\beta)}$ is a function of $x^a - x_{(\beta)}{}^a$]. In either case

$$\frac{d}{dt} \text{Tr}[I_{(\alpha)}(t) K_{(\beta)}] = 0 \;. \tag{8.47}$$

The interaction term in (8.46) involving particle positions is

$$\frac{1}{2k} \sum_{\alpha \neq \beta} \text{Tr}[I_{(\alpha)} K_{(\beta)}] \frac{d\psi_{(\beta)}(z_{(\alpha)})}{dt} \;. \tag{8.48}$$

While varying $x_{(\alpha)}$, we can treat $\text{Tr}[I_{(\alpha)}K_{(\beta)}]$ as constant in view of (8.47). Therefore, for variation of $x_{(\alpha)}$, (8.48) is effectively a total time derivative. hence $\ddot{x}_{(\alpha)} = 0$ as we require.

We must of course investigate if L_{eff} can be extended also to configurations with three or more overlapping $\Delta_{(\alpha)}$'s in such a way that the equations $\ddot{x}_{(\alpha)} = 0$ are still valid. It has been proved elsewhere [31] that the necessary conditions for the existence of such an extension are the "braid" quantization conditions. In quantum theory, they read

$$\exp\left\{\frac{i\pi}{k}\text{Tr}[\hat{I}_{(\alpha)}K_{(\gamma)}]\right\} \exp\left\{\frac{i\pi}{k}\text{Tr}[\hat{I}_{(\alpha)}K_{(\beta)}]\right\}$$
$$= \exp\left\{\frac{i\pi}{k}\text{Tr}[\hat{I}_{(\alpha)}K_{(\beta)}]\right\} \exp\left\{\frac{i\pi}{k}\text{Tr}[\hat{I}_{(\alpha)}K_{(\gamma)}]\right\},$$

α, β, γ distinct integers from the set $\{1, 2, \ldots, N\}$ (8.49)

$\hat{I}_{(\cdot)}$ denoting the quantum operator for the classical variable $I_{(\cdot)}$. It can be derived by considering relations such as: $\sigma_\alpha \sigma_{\alpha+1} \sigma_\alpha = \sigma_{\alpha+1} \sigma_\alpha \sigma_{\alpha+1}$ $[\alpha \leq N - 2]$ satisfied by the exchanges of particles α and $\alpha + 1$. These exchanges generate the braid group B_n.

Now the operator $\exp\{i\pi \text{Tr}[\hat{I}_{(\alpha)}K_{(\beta)}]/k\}$ represents the rotation by angle $2\pi j/k$ around the axis defined by $K_{(\beta)}$ in view of (830) and (8.38). It becomes $+1$ and -1 if

$$\frac{j}{k} \in \{0, \pm 1, \ldots\} \qquad (8.50)$$

which therefore is a particular solution of (8.49).

The first step in investigating the statistical properties of sources is to study the symmetry properties of L_{eff} under exchanges. L_{eff} is not always invariant under the naive exchanges $x_{(\alpha)} \to x_{(\beta)}$, $S_{(\alpha)} \to S_{(\beta)}$. We can however try to restore exchange symmetry by following up a naive exchange with internal transformations $S_{(\sigma)} \to T_{(\sigma)}^{-1} S_{(\sigma)}$, $T_{(\sigma)} \in \text{SU}(2)$ [see the remarks after (8.44) in this connection]. In order that L_{eff} is exchange ivnariant, $T_{(\sigma)}$'s must exist such that L_{eff} is invariant under the combination of both these transformations. Such $T_{(\sigma)}$ do not always exist.

Let us specialize to $N = 3$. In this case, the exchange can be written as

$$x_{(\alpha)} \to x_{(\beta)}, \qquad x_{(\gamma)} \to x_{(\gamma)}, \qquad \gamma \neq \alpha, \beta$$
$$S_{(\alpha)} \to G_{\alpha\beta}^{-1} S_{(\beta)}, \qquad S_{(\beta)} \to G_{\beta\alpha}^{-1} S_{(\alpha)}, \qquad S_{(\gamma)} \to H_{\alpha\beta}^{-1} S_{(\gamma)}. \qquad (8.51)$$

Here $G_{\alpha\beta}$, $G_{\beta\alpha}$ and $H_{\alpha\beta}$ must fulfill

$$G_{\alpha\beta} K_{(\beta)} G_{\alpha\beta}^{-1} = K_{(\alpha)}, \qquad G_{\alpha\beta} K_{(\gamma)} G_{\alpha\beta}^{-1} = K_{(\gamma)},$$
$$G_{\beta\alpha} K_{(\alpha)} G_{\beta\alpha}^{-1} = K_{(\beta)}, \qquad G_{\beta\alpha} K_{(\gamma)} G_{\beta\alpha}^{-1} = K_{(\gamma)}, \qquad (8.52)$$
$$H_{\alpha\beta} K_{(\beta)} H_{\alpha\beta}^{-1} = K_{(\alpha)}, \qquad H_{\alpha\beta} K_{(\alpha)} H_{\alpha\beta}^{-1} = K_{(\beta)}.$$

Evidently, we can assume that

$$G_{\beta\alpha} = G_{\alpha\beta}^{-1}, \qquad H_{\alpha\beta} = H_{\beta\alpha}. \qquad (8.53)$$

If such $G_{\alpha\beta}$, $H_{\alpha\beta}$ do not exist, L_{eff} is not exchange invariant.

It is very easy to choose $K_{(\alpha)}$'s so as to spoil the exchange invariance of L_{eff}. For if G_{12} for instance exists, then $\text{Tr}[K_{(1)}K_{(3)}] = \text{Tr}[K_{(2)}K_{(3)}]$ while if all $H_{\alpha\beta}$ exist, then $\text{Tr}[K_{(\alpha)}K_{(\beta)}]$ for $\alpha, \beta = 1, 2: 1, 3$ and $2, 3$ are equal. So all we have to do is to choose $K_{(\alpha)}$'s which violate these identities. One such choice is $K_{(1)} = j\tau_1$, $K_{(2)} = j\tau_2$, $K_{(3)} = j\tau_2$. There are also exchange invariant choices for $K_{(\alpha)}$, one such being $K_{(\alpha)} = j\tau_1$.

There is another way to see why L_{eff} may not always be exchange invariant. The potential seen by particle α is $A_{(\alpha)}$ [as given by (8.43) with $N = 3$]. We consider that the general definition of the exchange $\alpha \leftrightarrow \beta$ is of the form (8.51). Here the internal transformations must clearly be such that (8.51) effectively exchanges the potential $A_{(\alpha)}$ and $A_{(\beta)}$ in L_{eff} and leaves $A_{(\gamma)}$ invariant. This requires that $G_{\alpha\beta}$, $H_{\alpha\beta}$ fulfill (8.52). Hence when $G_{\alpha\beta}$, $H_{\alpha\beta}$ fail to exist, the potentials seen by the particles are not exchanged by any $\alpha \leftrightarrow \beta$ exchange regardless of its precise definition. The only restriction is that it transforms only the dynamical variables $x_{(\cdot)}$, $S_{(\cdot)}$ and does not transform $K_{(\cdot)}$.

When L_{eff} is not invariant under an exchange σ_α, σ_α transforms L_{eff} to a new Lagrangian $\sigma_\alpha L_{\text{eff}}$. All the Lagrangians L_{eff}, $\sigma_\alpha L_{\text{eff}}$, $\sigma_\alpha \sigma_\beta L_{\text{eff}}, \ldots$ must be considered together in a sense to be explained below to restore exchange invariance of the system and identity of the particles.

We may anticipate that the statistics of the sources [as defined by the representation of the braid group B_n on the quantum states; see Sec. (7.3)] depends on the symmetry properties of $A_{(\alpha)}$. This is correct and distinct B_n representations occur for suitable potential choices. The consequence is that the non-abelian sources are not uniquely associated with a particular statistics, rather the latter depends on the choice of $A_{(\alpha)}$.

Let us now consider exchange invariant L_{eff} for $N = 3$ and $j = 1/2$ compatible with (8.49).

The space of states for an exchange invariant L_{eff} is spanned by $|x_{(1)}x_{(2)}x_{(3)}\rangle \times |m_{(1)}m_{(2)}m_{(3)}\rangle$ where the internal group of particle α acts on the index $m_{(\alpha)}$ [$=1/2$ or $-1/2$]. It is important for our purposes to be explicit about the phase conventions involved in defining these states. Let $|x\rangle$ denote the usual single particle state localized at $x \in \mathbb{R}^2$ on which the single particle momenta p_α act as $-i\partial_\alpha$ [here we regard x as x_1, x_2]. Let $|m\rangle$ [$m = 1/2, -1/2$] denote the states which transform under $\Gamma^{(1/2)}$ by the standard $D^{(1/2)}$ matrices, m being the third component of angular momentum. Then $|x_{(1)}x_{(2)}x_{(3)}\rangle|m_{(1)}m_{(2)}m_{(3)}\rangle$ is the tensor product $\{\bigotimes_\alpha |x_{(\alpha)}\rangle\} \otimes \{\bigotimes_\alpha |m_{(\alpha)}\rangle\}$. It is ambiguous only by an overall constant phase.

The transformation of this basis under $\hat{\sigma}_\alpha$ is obtained by first applying the naive exchange $\hat{N}_\alpha$ which interchanges $x_{(\alpha)}$, $m_{(\alpha)}$ with $x_{(\alpha+1)}$, $m_{(\alpha+1)}$. For instance

$$\hat{N}_1 |x_{(1)}x_{(2)}x_{(3)}\rangle |m_{(1)}m_{(2)}m_{(3)}\rangle = |x_{(2)}x_{(1)}x_{(3)}\rangle |m_{(2)}m_{(1)}m_{(3)}\rangle. \qquad (8.54)$$

As the phase uncertainty in the basis states is only an overall constant phase, $\hat{N}_\alpha$ are uniquely defined by these equations. We cannot for instance replace $\hat{N}_\alpha$ by $i\hat{N}_\alpha$. The action of $\hat{\sigma}_\alpha$ on these states is obtained by following up the application of $\hat{N}_\alpha$ by internal transformations $\hat{G}_{\alpha+1,\alpha}$, $\hat{G}_{\alpha,\alpha+1}$, $\hat{H}_{\alpha,\alpha+1}$ associated with the previously discussed internal transformations on the $S_{(\cdot)}$'s. For $\alpha = 1$, for instance, they act on the first, second and third internal indices of the states. Thus $\hat{\sigma}_\alpha = \hat{H}_{\alpha,\alpha+1}\hat{G}_{\alpha,\alpha+1}\hat{G}_{\alpha+1,\alpha}\hat{N}_\alpha$ and $\hat{\sigma}_1$ for instance acts according to

$$\hat{\sigma}_1 |x_{(1)}x_{(2)}x_{(3)}\rangle |m_{(1)}m_{(2)}m_{(3)}\rangle = |x_{(2)}x_{(1)}x_{(3)}\rangle |lmn\rangle [G_{21}]_{lm_{(2)}}[G_{12}]_{mm_{(1)}}[H_{12}]_{nm_{(3)}}. \tag{8.55}$$

With this action and the choice $H_{23} = G_{21}G_{32}H_{12}G_{23}G_{12}$ [which is clearly compatible with (8.52)], we have $\hat{\sigma}_\alpha\hat{\sigma}_{\alpha+1}\hat{\sigma}_\alpha = \hat{\sigma}_{\alpha+1}\hat{\sigma}_\alpha\hat{\sigma}_{\alpha+1}$ so that $\hat{\sigma}_\alpha$'s generate a representation of B_3.

It must be decomposed into unitary irreducible representations (UIR's) to obtain states transforming irreducibly under B_3. In any case, using (8.53), we find

$$\hat{\sigma}_1^2 |x_{(1)}x_{(2)}x_{(3)}\rangle |m_{(1)}m_{(2)}m_{(3)}\rangle = |x_{(1)}x_{(2)}x_{(3)}\rangle |m_{(1)}m_{(2)}n\rangle [H_{12}^2]_{nm_{(3)}} \tag{8.56}$$

with similar equations for $\hat{\sigma}_2^2$ and $\hat{\sigma}_3^2$.

Note that for the choice $K_{(\alpha)} = \tau_1/2$, we can set $G_{\alpha\beta} = H_{\alpha\beta} = 1$ and $\hat{\sigma}_\alpha^2 = 1$.

We finally explain how to treat an exchange noninvariant L_{eff} for $N = 3$ and $j = 1/2$. It involves considering not only L_{eff}, but also all its distinct transforms generated by exchanges and their products. For one L_{eff} characterized by the triple $\mathbf{K} = (K_{(1)}, K_{(2)}, K_{(3)})$, the quantum states are assumed to be spanned by

$$|x_{(1)}x_{(2)}x_{(3)}\rangle |m_{(1)}m_{(2)}m_{(3)}\rangle |(K_{(2)}, K_{(3)}), (K_{(3)}, K_{(1)}), (K_{(1)}, K_{(2)})\rangle.$$

The factor $|(K_{(2)}, K_{(3)}), (K_{(3)}, K_{(1)}), (K_{(1)}, K_{(2)})\rangle$ has been inserted to emphasize that these are appropriate for this L_{eff} [note that this factor is characterized only by the three vectors $K_{(1)}, K_{(2)}, K_{(3)}$ indicating that we can denote it equally by $|K_{(1)}K_{(2)}K_{(3)}\rangle$]. There is also an abuse of notation here: these states are not tensor products, but "twisted" products obeying the identity.

$$|x_{(1)}x_{(2)}x_{(3)}\rangle |l_{(1)}l_{(2)}l_{(3)}\rangle |(K_{(2)}, K_{(3)}), (K_{(3)}, K_{(1)}), (K_{(1)}, K_{(2)})\rangle$$
$$\times g_{(1)l_{(1)}m_{(1)}}g_{(2)l_{(2)}m_{(2)}}g_{(3)l_{(3)}m_{(3)}} = \eta |x_{(1)}x_{(2)}x_{(3)}\rangle |m_{(1)}m_{(2)}m_{(3)}\rangle$$
$$\times |g_{(1)}^{-1}(K_{(2)}, K_{(3)})g_{(1)}, g_{(2)}^{-1}(K_{(3)}, K_{(1)})g_{(2)}, g_{(3)}^{-1}(K_{(1)}, K_{(2)})g_{(3)}\rangle \tag{8.57}$$

for $g_{(i)}$, $g_{(j)}$ which fulfill

$$g_{(i)}g_{(j)}^{-1}K_{(k)}(g_{(i)}g_{(j)}^{-1})^{-1} = K_{(k)}, \tag{8.58}$$

$i, j, k =$ any cyclic permutation of 1, 2, 3.

Here for example

$$g_{(1)}^{-1}(K_{(2)}, K_{(3)})g_{(1)} = (g_{(1)}^{-1}K_{(2)}g_{(1)}, g_{(1)}^{-1}K_{(3)}g_{(1)}) \,. \quad (8.59)$$

(8.57) can be understood by noting that the transformations $S_{(i)} \to g_{(i)}S_{(i)}$ where $g_{(i)}$ fulfill (8.58) are equivalent to the transformations

$$K_{(1)} \to g_{(2)}^{-1}K_{(1)}g_{(2)} \,, \quad K_{(2)} \to g_{(3)}^{-1}K_{(2)}g_{(3)} \,, \quad K_{(3)} \to g_{(1)}^{-1}K_{(3)}g_{(1)}$$

in L_{eff}. The constraints on $g_{(i)}g_{(j)}^{-1}$ in (8.58) come from requiring that the ket

$$|g_{(1)}^{-1}(K_{(2)}, K_{(3)})g_{(1)}, g_{(2)}^{-1}(K_{(3)}, K_{(1)})g_{(2)}, g_{(3)}^{-1}(K_{(1)}, K_{(2)})g_{(3)}\rangle$$

is characterized only by the three vectors $g_{(2)}^{-1}K_{(1)}g_{(2)}, g_{(3)}^{-1}K_{(2)}g_{(3)}, g_{(1)}^{-1}K_{(3)}g_{(1)}$. Finally η is a phase. It cannot always be set equal to 1. For $g_{(1)} = g_{(2)} = 1$, $g_{(3)} = -1$ for example it is -1.

Hereafter, we shall use the notation $|K_{(1)}K_{(2)}K_{(3)}\rangle$ to denote the ket $|(K_{(2)}, K_{(3)}), (K_{(3)}, K_{(1)}), (K_{(1)}, K_{(2)})\rangle$.

L_{eff} is invariant if the naive exchange $x_{(\alpha)}, S_{(\alpha)} \overset{N_g}{\leftrightarrow} x_{(\alpha+1)}$, is accompanied by the exchange $K_{(\alpha)} \leftrightarrow K_{(\alpha+1)}$. But since $K_{(\cdot)}$'s are not dynamical in L_{eff}, we must try to bring about their exchange by the internal transformations $\hat{G}_{\alpha+1,\alpha}, \hat{G}_{\alpha,\alpha+1}$ and $\hat{H}_{\alpha,\alpha+1}$ using (8.58)–(8.59) so long as we work only with a fixed L_{eff}. When this is possible, $\hat{\sigma}_\alpha$ transforms the states without affecting **K**. Suppose on the other hand that these internal transformations fail to exist, for instance, for the choice $\alpha = 1$. Then in order to maintain exchange symmetry of the sources, we are forced to enlarge the family of states $|x_{(1)}x_{(2)}x_{(3)}\rangle|m_{(1)}m_{(2)}m_{(3)}\rangle|K_{(1)}K_{(2)}K_{(3)}\rangle$ by adjoining the new state $|x_{(2)}x_{(1)}x_{(3)}\rangle|m_{(2)}m_{(1)}m_{(3)}\rangle|K_{(2)}K_{(1)}K_{(3)}\rangle$. The action of the exchange $\hat{\sigma}_1$ is then

$$\hat{\sigma}_1|x_{(1)}x_{(2)}x_{(3)}\rangle|m_{(1)}m_{(2)}m_{(3)}\rangle|K_{(1)}K_{(2)}K_{(3)}\rangle$$

$$= \eta_{(1)_k}|x_{(2)}x_{(1)}x_{(3)}\rangle|m_{(2)}m_{(1)}m_{(3)}\rangle|K_{(2)}K_{(1)}K_{(3)}\rangle \,, \quad (8.60)$$

where $\eta_{(1)_k}$ denotes a phase factor depending on the vector **K**. There is such a phase factor $\eta_{(2)_k}$ for $\hat{\sigma}_2$ as well [these phase factors, just as that occurring in (8.57), cannot always be set equal to 1].

For the choice (8.57), for example, in view of (8.60), $\eta_{(1)k_{(1)}k_{(2)}k_{(3)}}\eta_{(1)k_{(2)}k_{(1)}k_{(3)}} = -1$ with a similar equation for $\eta_{(2)_k}$. The new states also obey relations analogous to (8.57). There is such a family for every such exchange or product of exchanges although they may not all be distinct. This will be the case for two exchanges $\sigma_\alpha, \sigma_\beta$ if $\sigma_\alpha^{-1}\sigma_\beta$ can be implemented on the original family without affecting the arguments of kets involving $K_{(\cdot)}$'s.

The method for finding the UIR's of the group generated by exchanges can now be outlined. Let $\mathcal{H}_\mathbf{k}$ be the group generated by all those exchanges which can be implemented on the original vector space $V_\mathbf{k}$ of states [associated with L_{eff}] without changing it to a new vector space. The remaining exchanges change this vector space. now $\mathcal{H}_k$ can be regarded as a subgroup of B_3 so that a representation of B_3 can be found by the method of induced representations. The resulting B_3 representation is in general reducible. It can at least partially be reduced by reducing the $\mathcal{H}_k$ representation on V_k into UIR's and inducing a representation of B_3 starting from one of these UIR's.

The self-interaction of sources which has hitherto been ignored will now be briefly considered. It is enough to discuss one source, the result for N sources being obtained by addition. The following is proved in Ref. [31]. If the source is approached in angular direction ϕ, any solution of (8.35)–(8.36) [for the group SU(2) and for a general j and k] has then the limit

$$A(\phi) = -\frac{i}{k} I_{(\alpha)} d\Lambda(\phi) \tag{8.61}$$

where

$$\int_{\phi=0}^{2\pi} d\Lambda(\phi) = 2\pi . \tag{8.62}$$

With the choice $\Lambda(\phi) = \phi$, the self-interection term for one source is found from (8.31) to be

$$L_{\text{spin}} = \frac{j^2}{k} \dot{\phi}(t) . \tag{8.63}$$

This expression is to be interpreted in terms of a frame attached to the particle as in the discussion following (7.39). A rotation of this frame by an angle θ with particle position as centre changes ϕ to $\phi + \theta$. The momentum

$$p_\phi = \frac{j^2}{k} \tag{8.64}$$

conjugate to ϕ is hence the intrinsic spin of the particle. The particle therefore has intrinsic spin j^2/k. For $j = 1/2$ and $k = \pm 1$, it is $\pm 1/4$.

In the non-abelian case, as we have seen, there is no unique UIR of B_n we can unambiguously associate with the N particle states. As this UIR determines the statistics of a particle, there is no obvious correlation in this case between the spin p_ϕ of a particle and its statistics.

9. Anyon Superconductivity

Since the original suggestion of Anderson [18] on the RVB state and related developments on the possible existence of spinon excitations [117], there has been increasing interest in systems of particles with exotic statistics [130, 182, 185] and their relevance to high-T_c superconductivity [8, 19, 23, 26, 39, 40, 51, 55, 63, 69, 80, 81, 94, 95, 102, 114, 124, 147, 179, 181, 188, 193, 194].

More recently it has been argued by Laughlin and collaborators [69, 114, 124] that fractional statistics in itself can lead to superconductivity. It has been argued as well that fractional statistics excitations can arise in certain spin systems which are generically referred to as chiral spin liquids [179].

The suggestion that a system of anyons can lead to superconductivity has been investigated further [51, 69, 102, 116, 176, 178, 179], and it has been demonstrated recently [51, 69, 102] that indeed certain anyon systems do exhibit superconductivity. The superconductivity exhibited by these anyon systems is novel in the sense that an order parameter of BCS kind does not seem to exist in this case.

There has been further investigation of these anyon systems and studies at finite temperature show that there is a partial Meissner effect at finite temperature. Also, there is no phase transition as all physical quantities vary smoothly with the temperature [54, 98]. We will discuss this briefly at the end of this section.

The demonstration of the fact that the anyon systems studied in Refs. [51] and [102] are superconducting consists primarily in showing the existence of the Meissner effect. The systems studied in [51, 102] lead to essentially the same considerations in this respect. The nature of the superconducting phase was further elaborated in [51] by showing the existence of a Nambu-Goldstone mode. In the following subsections we will briefly review the models discussed in Refs. [51] and [102]. Similar models have recently been studied by several people [69, 116, 176, 178, 179]. In particular, Fetter *et al.* [69] have also shown the existence of the Meissner effect.

9.1. *Anyon gas systems*

We have seen in Sec. 7 that one way to incorporate the fractional statistics is to include a Chern-Simons gauge field at the Lagrangian level. The Lagranigan for an ideal gas of fractional statistics particles can thus be written as [51]:

$$\mathcal{L} = \sum_\alpha \left\{ \frac{m}{2} \dot{\mathbf{x}}_\alpha^2 + e[-a_0(\mathbf{x}_\alpha) + \dot{\mathbf{x}}_\alpha \cdot \mathbf{a}(\mathbf{x}_\alpha)] \right\} + \frac{\mu}{2} \int d^2x \, \epsilon^{\lambda\nu\rho} a_\lambda \partial_\nu a_\rho \,. \quad (9.1)$$

Here $\mathbf{x}_\alpha$ are the particle coordinates, and a_ν is the Chern-Simons gauge field. We know from the discussion of Sec. 7 that such a Lagrangian describes classically free particles, and that an important effect of the Chern-Simons gauge field is to implement fractional statistics quantum mechanically.

Variations with respect to a_λ give the field equations

$$ej_\lambda = \frac{\mu}{2} \epsilon_{\lambda\nu\rho} f^{\nu\rho} \quad (9.2)$$

where j_λ and $f_{\nu\rho}$ are standard point particle current and the field strength respectively. Integrating the zeroth component of the field equations (9.2) we get:

$$eN = \mu\Phi \qquad (9.3)$$

where N is the particle number and Φ is the flux corresponding to the Chern-Simons gauge field.

Note here that (9.2) implies that the field strength $f_{\nu\rho}$ is confined to the particle worldlines. This indicates that the effect of the Chern-Simons term is to associate with each particle a fictitious flux e/μ. Since the particles also carry a fictitious charge e, the wave functions will acquire phase through the Aharonov-Bohm effect as they wind around one another. This is the content of the fractional statistics. The angle θ characterizing the fractional statistics (see Sec. 7) can be easily calculated as

$$\theta = \frac{e^2}{2\mu}. \qquad (9.4)$$

Note that since the time derivative of a_0 never appears in (9.1), varying with respect to it simply yields the constraint

$$ej_0 = \mu\epsilon_{ij}\partial_i a_j \equiv \mu b \qquad (9.5)$$

where b is the magnetic field.

It was argued in Ref. [51] that a consistent model can be developed by considering the statistics of the anyons to be near the Fermi statistics. One may thus choose:

$$\theta = \pi\left(1 - \frac{1}{n}\right). \qquad (9.6)$$

This system can be investigated using the mean field approximation [27, 69]. This consists in replacing the total effect of distant particles by their average. In our context, this amounts to replacing the many flux tubes attached to the particles by a smooth magnetic field with the same total flux. For

$$\Delta\theta = \frac{\pi}{n}$$

one can relate the resulting magnetic field to the particle density ρ using (9.5)

$$b = \frac{2\pi}{en}\rho. \qquad (9.7)$$

The physical validity of the mean field approximation can be checked by the following considerations.

We know that in a uniform magnetic field given by (9.7) the particles will move along cyclotron orbits with radius

$$r = \frac{mv}{eb}. \tag{9.8}$$

Taking for v the velocity at the nominal Fermi surface, one can substitute

$$v = \sqrt{\frac{4\pi\rho}{m}}.$$

One finds that a typical cyclotron orbit contains

$$\rho\pi r^2 = n^2$$

particles on the average. As this number becomes large for large n, the mean field approximation can be expected to be a good approximation for large n.

A qualitative picture of the Meissner effect

Since in our mean field approximation the fermions are propagating in a uniform background fictitious magnetic field $b = (2\pi\rho/en)$, the energy eigenstates of the fermions will form Landau bands with degeneracy

$$\rho_l = \frac{eb}{2\pi} = \frac{\rho}{n} \tag{9.9}$$

per unit area and with energy eigenvalues (in natural units)

$$\epsilon_l = \left(l + \frac{1}{2}\right)\frac{eb}{m} \equiv \left(l + \frac{1}{2}\right)\omega_c \tag{9.10}$$

where $l = 0, 1, 2, \ldots$. When the statistical parameter θ is $\pi(1 - 1/n)$, the density is just such as to fill n Landau levels exactly.

Now in order to see if there is a Meissner effect present, let us consider the effect of adding a small real magnetic field B to the fictitious one b.

First let us consider the case when B and b are in the same direction. Then the density of states per Landau level will be somewhat greater, and we will not quite completely fill n levels anymore. By denoting the fractional filling of the highest level by $1 - x$, one can use the conservation of particle number to get

$$(b + B)(n - x) = bn,$$
$$(b + B)x = Bn. \tag{9.11}$$

The total energy then becomes

$$E = \frac{e(b + B)}{2\pi} \frac{e(b + B)}{m} \left\{ \sum_{l=0}^{n-1} \left(l + \frac{1}{2}\right) - \left(n - \frac{1}{2}\right) x \right\}$$

$$= \frac{n^2 e^2}{4\pi m} \left\{ b^2 + \frac{bB}{n} - \left(1 - \frac{1}{n}\right) B^2 \right\} . \qquad (9.12)$$

Thus, the energy relative to the ground state is positive and grows linearly with B for small B.

Now let us consider the case when B and b have opposite directions. In this case the density of states per Landau level will be smaller and some particles will have to occupy the $(n + 1)$st level. Denoting the fractional filling of this level again by x, we have from particle conservation:

$$(b - B)(n + x) = bn ,$$
$$(b - B)x = Bn . \qquad (9.13)$$

The energy then becomes

$$E = \frac{e(b - B)}{2\pi} \frac{e(b - B)}{m} \left\{ \sum_{l=0}^{n-1} \left(l + \frac{1}{2}\right) + \left(n + \frac{1}{2}\right) x \right\}$$

$$= \frac{n^2 e^2}{4\pi m} \left\{ b^2 + \frac{bB}{n} - \left(1 + \frac{1}{n}\right) B^2 \right\} . \qquad (9.14)$$

Comparing (9.12) and (9.14) one sees that the leading behavior in B for small B is the same in both cases.

The above considerations suggest that the anyon gas will tend to exclude the external magnetic fields and hence there is a kind of Meissner effect present in this system.

The order parameter and the massless mode

A central feature of superconductivity is the existence of a Nambu-Goldstone mode, or concretely an excitation with dispersion relation $\omega^2 \propto k^2$ at low frequency and small wavevector. For an anyon system, the existence of such a mode was shown first by Fetter, Hanna and Laughlin [69]. They calculated the effect of adding back the residual interactions, and found that these interactions produced the necessary pole in the current-current correlation function. Physically this means that there are particle-hole bound states at zero energy. Following a different approach, Chen, Wilczek, Witten and Halperin [51] showed the existence of the massless mode by showing that the commutativity of the momentum generators is spontaneously violated. In the following, we will present a brief discussion of this latter approach as it also exhibits the nature of the order parameter.

As we mentioned above, the main point in the approach in Ref. [51] is that while microscopically the momentum generators commute

$$[P_i, P_j] = 0 \tag{9.15}$$

macroscopically, at the level of quasiparticles, one obtains

$$[P_i, P_j] = if\epsilon_{ij}\mathcal{Q} \tag{9.16}$$

where $\mathcal{Q}$ is the particle number and f is a constant which can be interpreted as the fundamental order parameter of the anyon gas. We will see in what follows that there must be a massless spin zero boson associated with the spontaneous breaking of (9.15).

In order to see why (9.16) is true, let us note that if the P_i's are conserved and commute then it must be possible to take the quasiparticle excitations to be momentum eigenstates. This is however not possible, since the charged quasiparticles are not in plane wave states, but are in Landau levels (as we showed earlier). Now the charged quasiparticles are not plane waves due to the interaction with a nonzero expectation value of the fictitious magnetic field

$$f = \frac{1}{2}\epsilon^{ij}(\partial_i a_j - \partial_j a_i) .$$

Since we take f to be translationally invariant, the conservation of the P_i's is not spontaneously broken. On the other hand, in a magnetic field the translation generators do not commute. Hence, it is the commutativity of the P_i's which is spontaneously violated due to the nonzero expectation value of f.

To be specific, let us first consider a single particle moving in a constant magnetic field. The one-particle Hamiltonian is

$$\mathcal{H} = -\frac{1}{2m}\sum_i D_i^{(0)2} \tag{9.17}$$

where the covariant derivatives $D_i^{(0)}$ obey

$$[D_i^{(0)}, D_j^{(0)}] = i\epsilon_{ij}f . \tag{9.18}$$

The superscript "(0)" indicates that the gauge field is not dynamical. Note that the translation generators are not simply the covariant derivatives $D_i^{(0)}$, since these do not commute with $\mathcal{H}$. The conserved translation generators are

$$P_i = -iD_i^{(0)} + \epsilon_{ij}fx^j . \tag{9.19}$$

It is easy to see that the P_i's commute with $\mathcal{H}$ but do not commute with each other

$$[P_i, P_j] = if\epsilon_{ij} . \tag{9.20}$$

To see this in a second quantized formalism, let us recall that in studying the anyon gas we found an expectation value of the fictitious magnetic field f, quasiparticles that can occupy all Landau levels but the first n and quasiholes that can fill any state in the first n Landau levels. These quasiparticles and quasiholes can be represented by an effective fermion field χ with action

$$S = \int dt d^2x \left\{ \chi^* \left(i \frac{D^{(0)}}{Dt} \right) \chi - \frac{1}{2m} (D_k^{(0)} \chi^*)(D_k^{(0)} \chi) \right\}. \quad (9.21)$$

Here the gauge field is not dynamical. It should be noted that (9.21) is not a microscopic Lagrangian but (a piece of) a phenomenological Lagrangian in which as much as possible of the relevant physics is visible at tree level.

The Hamiltonian derived from (9.21) is

$$\mathcal{H} = \frac{1}{2m} \int d^2x D_k^{(0)} \chi^* D_k^{(0)} \chi. \quad (9.22)$$

Following the one-particle result (9.19) one can see that the translation generators which commute with $\mathcal{H}$ are

$$P_i^{(x)} = \int d^2x \{ \chi^*(-iD_i^{(0)})\chi + f\epsilon_{ij} x^j \chi^* \chi \}. \quad (9.23)$$

The generators obey the following commutation relations

$$[P_i^{(x)}, P_j^{(x)}] = if\epsilon_{ij} \mathcal{Q} \quad (9.24)$$

where

$$\mathcal{Q} = \int d^2x \chi^* \chi \quad (9.25)$$

is the conserved charge operator. (9.24) is the second-quantized version of the single-particle result (9.20).

Now we can see that the quasiparticle and the quasiholes (that are visible in lowest order in $1/n$) cannot be the whole story. This is because at a microscopic level the translation generators of the anyon gas commute but, in the phenomenological model (9.21), the translation generators do not commute.

We will see in the following that if in addition to the quasiparticles and quasiholes described in (9.21) we assume the existence of an additional spin zero massless boson, the above problem can be solved. This massless boson is analogous to a Goldstone boson associated with the spontaneous violation of the commutativity of the P_i's.

For our purposes it is convenient to represent this massless spin zero boson by an abelian gauge field b_i with field strength

$$h_{ij} = \partial_i b_j - \partial_j b_i \,.$$

The Lagrangian for b_i is

$$\mathcal{L}_b = \frac{1}{2} \int d^2x \left\{ \sum_j h_{0i}^2 - v^2 h_{12}^2 \right\} . \tag{9.26}$$

Here v is the velocity of propagation of the massless boson. This description is possible in 2+1 dimensions. A more conventional representation of the massless spin zero boson can be obtained by the change of variables: $\partial_0 \phi = h_{12}$, $v^2 \partial_i \phi = \epsilon_{ij} h_{0j}$. with this, (9.26) becomes

$$\mathcal{L} = -v^2 \int d^2x \left\{ \frac{1}{2} (\partial_0 \phi)^2 - \frac{v^2}{2} \sum_i (\partial_i \phi)^2 \right\} .$$

The translation generator of an abelian gauge field is

$$P_i = \int d^2x T^{(b)}{}_{0i} \tag{9.27}$$

where the conventional form of the momentum density is

$$T^{(b)}{}_{0i} = -\sum_j h_{0j} h_{ij} \,. \tag{9.28}$$

This leads to the standard result $[P_i, P_j] = 0$. However, by adding a term to (9.28)

$$T^{(b)}{}_{0i} = -\sum_j h_{0j} h_{ij} + \epsilon_{ij} f h_{0j} \tag{9.29}$$

one can see, using the canonical commutation relations derived from (9.29), that

$$[P^{(b)}{}_i, P^{(b)}{}_j] = if\epsilon_{ij} \int d^2x [\partial_k h_{0k}] = if\epsilon_{ij} \int_{\mathscr{C}} dl n^k h_{0k} \tag{9.30}$$

where the integral is over a large circle $\mathscr{C}$ at infinity, and n^k is the normal vector to the circle.

Now if we combine the χ and b systems and form the total momentum operator $P_i = P_i^{(\chi)} + P_i^{(b)}$ then we get

$$[P_i, P_j] = if\epsilon_{ij} \left(\mathcal{Q} - \int_{\mathscr{C}} dl n^k h_{0k} \right) . \tag{9.31}$$

Thus the commutativity of the P_i's can be restored by restricting to the subspace of the Hilbert space for which

$$\mathcal{Q} - \int_{\mathscr{C}} dl n^k h_{0k} = 0 \ . \tag{9.32}$$

However, if we just consider (9.21) and (9.26), then the Gauss's law constraint implies $\partial_k h_{0k} = 0$, and hence

$$\int_{\mathscr{C}} dl n^k h_{0k} = 0$$

which is in contradiction with (9.32). This problem can be solved by modifying the free Lagrangian by adding a suitable coupling of b to χ. The requisite Lagrangian is

$$\mathcal{L} = \frac{1}{2} \int d^2x (h_{0i}{}^2 - v^2 h_{12}{}^2) + \int d^2x \left(\chi^* \left(i \frac{D}{Dt} \right) \chi - \frac{1}{2m} D_k \chi^* D_k \chi \right) \tag{9.33}$$

where now: $D_i = \partial_i + ib_i$, and b_i is a dynamical gauge field.

Now one can see the role of b_i (the massless spin zero boson) in restoring the commutativity of the translations. (9.33) is a perfectly normal Lagrangian with commuting translation generators. By expanding around a constant expectation value f of the "magnetic field" $\partial_1 b_2 - \partial_2 b_1$, one will find the χ excitations to be Landau orbits with apparent violation of commutativity of translations. But this phenomenon is "spontaneous" in the sense that is simply reflects the non-zero expectation value of h.

This concludes the demonstration of the existence of a massless mode for the anyon gas.

9.2. *A field-theoretic system of electrons and the Chern-Simons field*

Superconductivity due to the presence of a Chern-Simons field was also demonstrated (by following a somewhat different approach) by Hosotani and Chakravarty [102] by considering a field-theoretic system consisting of the electron (or hole) fields ψ_σ, the electromagnetic gauge field A_μ and a Chern-Simons gauge field a_μ. The Lagrangian they considered is as follows

$$\mathcal{L} = -\frac{1}{4} F_{\mu\nu}{}^2 - \frac{\mu}{2} \epsilon^{\lambda\nu\rho} a_\lambda \partial_\nu a_\rho + \psi^\dagger (iD_0) \psi - \frac{1}{2m} |D_k \psi|^2 - \frac{e}{2m} B \psi^\dagger \sigma_3 \psi - eA_0 n_e \tag{9.34}$$

where $D_\mu = \partial_\mu - ie(A_\mu + a_\mu)$, $B = F_{12}$, $b = f_{12}$. n_e is the electron (or hole) density and $F_{\mu\nu}$ ($f_{\mu\nu}$) are the field strengths. The last term accounts for the background neutralizing charges.

The equations of motion were derived in the mean field approximation as

$$\frac{1}{2} \mu \epsilon^{\lambda\nu\rho} f_{\nu\rho} = \langle j^\lambda \rangle$$

$$\partial_\nu F^{\lambda\nu} = \langle J^\lambda \rangle - e n_e \delta^{\lambda 0} \tag{9.35}$$

where

$$j^0 = J^0 = e\psi^\dagger\psi$$

$$j^k = -\frac{ie}{2m}(\psi^\dagger(D_k\psi) - (D_k\psi^\dagger)\psi) = J^k - \frac{e}{2m}\epsilon^{kj}\partial_j(\psi^\dagger\sigma_3\psi) \tag{9.36}$$

and $\langle j^\mu \rangle$ is the expectation value of j^μ: $\langle\psi|j^\mu|\psi\rangle$.

Meissner effect

In the presence of weak external magnetic fields one can calculate the perturbed values of $\langle j^\lambda \rangle$. From here on we will choose: $|\mu| = e^2N/2\pi$ ($N = 2, 4, 6, \ldots$) for which case a complete Meissner effect will be derived. By writing $b = b^{(0)} + b^{(1)}$, where $b^{(1)}$ is the perturbation due to the effect of the external magnetic field, one gets [102]

$$\langle J^0 \rangle - en_e = \frac{Ne^2}{2\pi}(b^{(1)} + B) + \frac{N^2e^2m}{4\pi^2 n_e}\partial_2(f_{02} + E_2) + \cdots$$

$$\langle J^1 \rangle = -\frac{Ne^2}{2\pi}(f_{02} + E_2) - \frac{N^2e^2}{4\pi m}\partial_2(b^{(1)} + B) + \cdots . \tag{9.37}$$

Now it follows from (9.35) that

$$\langle J^0 \rangle - en_e = -\partial_2 E_2 = \frac{Ne^2}{2\pi}b^{(1)}$$

$$\langle J^1 \rangle = \partial_2 B = -\frac{Ne^2}{2\pi}f_{02}$$

for the case when the boundary of the superconductor lies along the 1 axis. Eliminating $b^{(1)}$ and f_{02} from (9.37) using the above equations and discarding the last term in the expression for $\langle J^1 \rangle$ in (9.37) (which is small for $N \ll 1000$) one gets the equation

$$\frac{Ne^2}{2\pi}\left\{B - \frac{m}{e^2 n_e}\partial_2^2 B\right\} = 0 \tag{9.38}$$

giving the Meissner effect explicitly. The penetration depth λ ($B \propto \exp[-x_2/\lambda]$) is given by

$$\lambda = \sqrt{\frac{m}{e^2 n_e}}. \tag{9.39}$$

It is interesting to note that although the mechanism for superconductivity is completely different, the expression for the penetration depth is exactly the same as in the BCS theory.

Similar conclusions regarding the Meissner effect have been obtained by Fetter *et al.* [69].

Meissner effect at finite temperatures

A finite temperature study of the above system has been carried out by Daemi, Salam and Strathdee [54] and by Hetrick, Hosotani and Lee [98a]. These results show that the penetration depth characterizing the exponential decay of the magnetic field inside the superconductor decreases with increasing temperature. Also, there is no indication of a phase transition.

It was further shown by Hetrick *et al.* in [98] that there is a critical magnetic field H_c (which depends on temperature) such that for external magnetic field $B_{ext} > H_c$, there is a partial penetration of the magnetic field inside the material. The value of the magnetic field inside the superconductor is then given by: $B_{in} = B_{ext} - H_c$. The magnetic field B near the boundary of a sample decreases from B_{ext} to B_{in} according to an exponential law with a penetration depth λ_0 which is the same penetration depth which was found to decrease with increasing temperature. λ_0 decreases slowly for $T < 70$ K, but rapidly approaches zero near $T = 100$ K. It was however argued in [98] that the penetration depth which is experimentally observed is not λ_0, but instead an effective penetration depth $\lambda_{eff} \sim B[dB/dx]^{-1}$. λ_{eff} increases slowly initially, but grows rapidly at temperature near about 100 K. However, λ_{eff} does not diverge at any temperature. There is no phase transition in a precise sense, since all physical quantities are smooth functions of T, but one can define a sort of critical temperature (about 100 K for the parameter values of Ref. [98]) above which superconductivity is effectively lost. For example, above this temperature even the partial Meissner effect completely vanishes and there are no supercurrents.

As has been mentioned in Ref. [98], the treatment in Refs. [54, 98] does not take into account the vortex-antivortex excitations which are expected to play an important role in the presence of an external magnetic field. Hetrick and Hosotani [98c] have computed the Hall voltage as a function of temperature. They find that the Hall voltage is initially extremely small for $T < T_c$ (where T_c is defined in the sense described above), increases rapidly for $T \sim T_c$, and then drops again to a very small value for $T > T_c$. We refer to Ref. [98c] for further details.

Acknowledgments

One of us (G. M.) would like to thank INFN for financial support and the physics department of the University of Syracuse (N.Y.) for the hospitality extended to him during the periods in which this work was initiated and brought to an end.

G. M. and E. E. would like to thank Prof. G. Fano for useful discussions and suggestions.

A. M. S. would lik to thank Y. Hosotani, A. J. Leggett, B. H. Lee and J. Hetrick for useful discussions. He also thanks S. Varma for her help in making corrections.

The work of A. P. B. and of A. M. S. was supported by the Department of Energy under contract number DE-FG02-85ER40231, and the work of A.M.S. was in addition supported by the Theoretical Physics Institute of the University of Minnesota.

A. P. B. also thanks the Centre for Theoretical Studies, Indian Institute of Science, Bangalore, India, for hospitality and support while this work was in progress.

References

[1] M. Abramowitz and I. A. Stegun, *Handbook of Mathematical Functions* (Dover, 1970), §23.
[2] A. A. Abrikosov, L. P. Gor'Kov and I. E. Dzyaloshinskii, *Methods of Quantum Field Theory in Statistical Physics*, (Prentice-Hall, N.Y., 1963).
[3] M. Acquarone, "Hubbard correlations in a single band", in *Physics of Metals*, eds. E. S. Giuliano and C. Rizzuto, (World Scientific, 1988) and references therein.
[4] I. Affleck, *Nucl. Phys.* **B257** (1985) 397.
[5] I. Affleck, *Nucl. Phys.* **B265** (1986) 409.
[6] I. Affleck and J. Brad Marston, *Phys. Rev.* **B37** (1988) 3744.
[7] I. Affleck and F. D. M. Haldane, *Phys. Rev.* **B36** (1987) 5291.
[8] I. Affleck, Z. Zou, T. Hsu and P. W. Anderson, *Phys. Rev.* **B38** (1988) 745.
[9] Y. Aharonov and D. Bohm, *Phys. Rev.* **115** (1959) 485.
[10] I. J. R. Aitchison and N. E. Mavromatos, *Phys. Rev.* **B39** (1989) 6544.
[11] I. J. R. Aitchison and N. E. Mavromatos, *Mod. Phys. Lett.* **A4** (1989) 521.
[12] I. J. R. Aitchison and N. E. Mavromatos, *Phys. Rev. Lett.* **63** (1989) 2684.
[13] M. G. Alford and F. Wilczek, Harvard Preprint HUTP-88/A047, 1988.
[14] P. W. Anderson, *Phys. Rev.* **115** (1959) 2.
[15] P. W. Anderson, *Phys. Rev.* **141** (1961) 41.
[16] P. W. Anderson, *Solid State Phys.* **14** (1963) 99.
[17] P. W. Anderson, *Mat. Res. Bull.* **8** (1973) 153.
[18] P. W. Anderson, *Science* **235** (1987) 1196.
[19] P. W. Anderson, "50 years of the Mott phenomenon", in *Frontiers and Borderlines in Many-Particle Physics*, Proc. of the Varenna Summer School (North-Holland, 1987).
[20] P. W. Anderson, "Some new ideas on RVB states: generalized flux phases", in *Physics of Low-dimensional Systems*, Proc. of Nobel Symposium 73, eds. S. Lundqvist and N. R. Nilsson (World Scientific, 1989).
[21] P. W. Anderson, G. Baskarans, Z. Zou and T. Hsu, *Phys. Rev. Lett.* **58** (1987) 2790.
[22] N. Andrei, K. Fukuya and J. H. Lowenstein, *Rev. Mod. Phys.* **55** (1983) 331.
[23] C. Aneziris, A. P. Balachandran and A. M. Srivastava, *Mod. Phys. Lett.* **B4** (1990) 439.
[24] C. Aneziris, A. P. Balachandran, M. Bourdeau, S. Jo, R. D. Sorkin and T. R. Ramadas, *Int. J. Mod. Phys.* **A4** (1989) 5495.
[25] D. P. Arovas, A. Auerbach and F. D. M. Haldane, *Phys. Rev. Lett.* **60** (1988) 531.
[26] D. P. Arovas, J. R. Schrieffer and F. Wilczek, *Phys. Rev. Lett.* **53** (1984) 722.
[27] D. P. Arovas, J. R. Schrieffer, F. Wilczek and A. Zee, *Nucl. Phys.* **B251** (1985) 117.
[28] A. P. Balachandran, "Skyrmions" in *High Energy Physics 1985*, Proc. of the Yale Theoretical Physics Institute, eds. M. J. Bowick and F. Gürsey, (World Scientific, 1986).
[29] A. P. Balachandran, "Classical topology and quantum phases", in *Anomalies, Phases, Defects . . .* , eds. M. Bregola, G. Marmo and G. Morandi, (Bibliopolis, Naples, 1990).
[30] A. P. Balachandran, M. Bourdeau and S. Jo, *Mod. Phys. Lett.* **A4** (1989) 1923.
[31] A. P. Balachandran, M. Bourdeau and S. Jo, Syracuse University Preprint SU-4228-406, 1989, and *Int. J. Mod. Phys*, in press.
[32] A. P. Balachandran, M. J. Bowick, K. S. Gupta and A. M. Srivastava, *Mod. Phys. Lett.* **A3** (1988) 1725.
[33] A. P. Balachandran, H. Gomm and R. Sorkin, *Nucl. Phys.* **B281** (1987) 573.
[34] A. P. Balachandran, G. Landi and B. Rai, Syracuse University Preprint SU-4228-423, 1989.
[35] A. P. Balachandran, B. Rai, G. Sparano and A. M. Srivastava, *Phys. Rev. Lett.* **59** (1987) 853.
[36] A. P. Balachandran, B. Rai, G. Sparano and A. M. Srivastava, *Int. J. Mod. Phys.* **A3** (1988) 2621.
[37] A. P. Balachandran, A. Stern and G. Trahern, *Phys. Rev.* **D19** (1979) 2416.
[38] G. Baskaran, *Int. J. Mod. Phys.* **B1** (1987) 697.

[39] G. Baskaran and P. W. Anderson, *Phys. Rev.* **B37** (1988) 580.
[40] G. Baskaran and R. Shankar, *Mod. Phys. Lett.* **B2** (1988) 1211.
[41] G. Baskaran, Z. Zou and P. W. Anderson, *Solid State Commun.* **63** (1987) 973.
[42] J. G. Bednorz and K. A. Müller, *Z. Phys.* **B64** (1986) 188.
[43] S. T. Beliaev, "Introduction to the Bogoliubov Transformation Method", in *The Many-Body Problem*, Les Houches, 1958. See also: G. Fano and G. Loupias, *Commun. Math. Phys.* **20** (1971) 143.
[44] H. A. Bethe, *Z. Phys.* **71** (1931) 205. A readable account of the Bethe Ansatz can be found in Refs. 134 and 171.
[45] J. Birman, "Braids, Links and Mapping Class Groups", *Annals of Math. Studies* **82**, (Princeton University Press, 1973).
[46] M. J. Bowick, P. Karabali and L. C. R. Wijewardhana, *Nucl. Phys.* **B271** (1986) 417.
[47] J. Brad Marston, *Phys. Rev. Lett.* **17** (1988) 1914.
[48] P. Chandra and B. Doucot, Princeton University Preprint, 1990.
[49] K. A. Chao, J. Spalek and A. M. Oles, *J. Phys.* **C10** (1977) L271.
[50] K. A. Chao, J. Spalek and A. M. Oles, *Phys. Rev.* **B18** (1978) 3453.
[51] Y. H. Chen, F. Wilczek, E. Witten and B. Halperin, *Int. J. Mod. Phys.* **B3** (1989) 1001.
[52] M. Cyrot, *J. Phys. (Paris)* **33** (1972) 125.
[53] A. D'adda, M. Luscher and P. Di Vecchia, *Phys. Rep.* **49C** (1979) 239.
[54] S. R. Daemi, A. Salam and J. Strathdee, ICTP Preprint IC/89/283, 1989.
[55] E. Dagotto, E. Fradkin and A. Moreo, *Phys. Rev.* **B38** (1988) 2926.
[56] S. Deser, R. Jackiw and S. Templeton, *Ann. Phys. (N.Y.)* **140** (1982) 372.
[57] S. Deser, R. Jackiw and S. Templeton, *Ann. Phys. (NY.)* **140** (1982) 420.
[58] S. Deser, R. Jackiw and S. Templeton, *Phys. Rev. Lett.* **48** (1982) 975.
[59] Ph. De Sousa Gerbert, MIT Preprint CTP #1653, 1988.
[60] Ph. De Sousa Gerbert and R. Jackiw, MIT Preprint, 1988.
[61] A. M. Din and W. J. Zakrzewski, *Nucl. Phys.* **B253** (1985) 77.
[62] G. V. Dunne, R. Jackiw and C. A. Trugenberger, MIT Preprint CTP-1711, 1989.
[63] L. Dzyaloshinskii, A. Polyakov and P. B. Wiegmann, *Phys. Lett.* **A127** (1988) 112.
[64] S. Elitzur, *Phys. Rev.* **D12** (1975) 3978. See also Ref. 118.
[65] V. J. Emery, *Phys. Rev.* **B14** (1976) 2989.
[66] E. Ercolessi, Thesis, Bologna 1989, unpublished.
[67] E. Ercolessi and G. Morandi, in preparation.
[68] P. Fazekas and P. W. Anderson, *Phil. Mag.* **30** (1974) 432.
[69] A. L. Fetter, C. B. Hanna and R. B. Laughlin, *Phys. Rev.* **B39** (1989) 9679.
[70] A. L. Fetter and J. D. Walecka, *Quantum Theory of Many-Particle Systems*, (McGraw-Hill, 1971).
[71] R. Floreanini, R. Percacci and E. Sezgin, ICTP Preprint IC/88/363, 1988, and references therein.
[72] E. Fradkin and M. Stone, Preprint ILL-(TH)-88#12.
[73] J. L. Friedman and R. Sorkin, *Phys. Rev. Lett.* **44** (1980) 1100.
[74] J. L. Friedman and R. Sorkin, *Phys. Rev. Lett.* **45** (1980) 148.
[75] J. L. Friedman and R. Sorkin, *Gen. Relativ. Grav.* **14** (1982) 615.
[76] J. Frölich and P. A. Marchetti, *Lett. Math. Phys.* **16** (1988) 347.
[77] J. Frölich and P. A. Marchetti, *Commun. Math. Phys.* **121** (1989) 177.
[78] T. H. Geballe and J. C. Hulm, *Science* **239** (1988) 367.
[79] A. S. Goldhaber, R. MacKenzie and F. Wilczek, Harvard Preprint HUTP-88/A044, 1988.
[80] T. R. Govindarajan and R. Shankar, Institute of Mathemtical Sciences Preprint, Madras 1988.
[81] M. Greiter, F. Wilczek and E. Witten, *Mod. Phys. Lett.* **B3** (1989) 903.
[82] C. Gros, R. Joynt and T. M. Rice, *Phys. Rev.* **B36** (1987) 381.
[83] J. Grundberg, T. h. Hansson, A. Karlhede and V. Lindstrom, *Phys. Lett.* **B218** (1989) 321.

[84] E. Guadagnini, M. Martellini and M. Mintchev, CERN and Pisa Preprint, CERN-TH 5419/89 (IFUP-TH 25/89), and references therein.
[85] E. Guadagnini, M. Martellini and M. Mintchev, CERN and Pisa Preprint, CERN-TH 5420/89 (IFUP-TH 24/89) and references therein.
[86] K. S. Gupta, G. Landi and B. Rai, *Mod. Phys. Lett.* **A4** (1989) 2263.
[87] F. Gürsey, *Nuovo Cim.* **7** (1958) 411.
[88] M. Gutzwiller, *Phys. Rev.* **A137** (1965) 1726.
[89] C. R. Hagen, Rochester Preprint 1989, and references therein.
[90] C. R. Hagen, *Ann. Phys. (N.Y.)* **157** (1984) 342.
[91] F. D. M. Haldane, *Phys. Lett.* **A93** (1983) 464.
[92] F. D. M. Haldane, *Phys. Rev. Lett.* **50** (1983) 1153.
[93] F. D. M. Haldane, *Phys. Rev. Lett.* **61** (1988) 1029.
[94] F. D. M. Haldane, *Phys. Rev. Lett.* **52** (1984) 1583.
[95] B. I. Halperin, J. March-Russell and F. Wilczek, Harvard Preprint HUTP-89/A010, 1989.
[96] A. J. Heeger, S. Kivelson and J. R. Schrieffer, *Rev. Mod. Phys.* **60** (1988) 781.
[97] C. Herring in *Magnetism*, Vol. IV, eds. G. T. Rado and H. Suhl, (Academic Press, N.Y., 1966).
[98] a) J. Hetrick, Y. Hosotani and B. H. Lee, University of Minnesota Preprint UMN-TH-831/90, TPI-MINN-90/40-T; 1990. See also: b) J. Hetrick, Ph.D Thesis, University of Minnesota, 1990, and c) J. Hetrick and Y. Hosotani, University of Minnesota preprint (under preparation).
[99] J. E. Hirsch, *Phys. Rev.* **B28** (1983) 4059.
[100] J. E. Hirsch, *Phys. Rev.* **B31** (1985) 4403.
[101] P. A. Horvathy, G. Morandi and E. C. G. Sudarshan, *Nuovo Cim.* **D11** (1989) 201.
[102] Y. Hosotani, Princeton Preprint, 1989; Y. Hosotani and S. Chakravarty, Princeton Preprint IASSNS-HEP-89/31, 1989. See also P. K. Panighrai, B. Ray and B. Sakita, City University of New York Preprint CCNY-HEP-89/22, 1989.
[103] C. Y. Huang, *Int. J. Mod. Phys.* **B2** (1988) 355.
[104] K. Huang and E. Manousakis, *Phys. Rev.* **B36** (1987) 8302.
[105] J. Hubbard, *Phys. Rev. Lett.* **3** (1959) 77.
[106] J. Hubbard, *Proc. Roy. Soc.* **A276** (1963) 238.
[107] J. Hubbard, *Proc. Roy. Soc.* **A277** (1964) 237.
[108] J. Hubbard, *Proc. Roy. Soc.* **A281** (1964) 401.
[109] D. H. Huse, *Phys. Rev.* **B37** (1988) 2380.
[110] L. B. Ioffe and I. Larkin, *Int. J. Mod. Phys.* **B2** (1988) 203.
[111] C. J. Isham, in *Relativity, Groups and Topology II*, eds. B. S. DeWitt and R. Stora (North-Holland, 1984).
[112] R. Jackiw and S. Templeton, *Phys. Rev.* **D23** (1981) 2291.
[113] P. Jain, J. Schechter and R. Sorkin, *Phys. Rev.* **D39** (1989) 998.
[114] V. Kalmeyer and R. B. Laughlin, *Phys. Rev. Lett.* **59** (1987) 2095.
[115] T. Kato, *Prog. Theor. Phys.* **4** (1959) 154.
[116] Y. Kitazawa and M. Murayama, University of Tokyo Preprint UT-549, 1989.
[117] S. A. Kivelson, D. S. Rokhsar and J. P. Sethna, *Phys. Rev.* **B35** (1987) 8865.
[118] J. Kogut, *Rev. Mod. Phys.* **51** (1979) 659.
[119] J. Kogut and L. Susskind, *Phys. Rev.* **D11** (1975) 395.
[120] G. Kotliar, *Phys. Rev.* **B37** (1988) 3664.
[121] M. G. Laidlaw and C. Morette-DeWitt, *Phys. Rev.* **D3** (1971) 1375.
[122] L. D. Landau and E. F. Lifshitz, *Statistical Physics Vol. V* (Pergamon Press, 1958).
[123] R. B. Laughlin, in *The Quantum Hall Effect, 2nd edition*, eds. R. E. Prange and S. M. Girvin, (Springer, 1990).
[124] R. B. Laughlin, *Science* **242** (1988) 525.
[125] R. B. Laughlin, *Phys. Rev. Lett.* **60** (1988) 2677.

[126] J. M. Leinnas and J. Myrheim, *Nuovo Cim.* **B37** (1977) 1.
[127] E. H. Lieb and F. Y. Wu, *Phys. Rev Lett.* **20** (1976) 1445.
[128] R. MacKenzie and F. Wilczek, *Int. J. Mod. Phys.* **A3** (1988) 2827.
[129] R. MacKenzie and F. Wilczek, Preprint DOE/ER/01545-406, 1989.
[130] R. MacKenzie, F. Wilczek and A. Zee, *Phys. Rev. Lett.* **53** (1984) 2203.
[131] E. Manousakis and R. Salvador, *Phys. Rev. Lett.* **60** (1988) 840.
[132] E. Manousakis and R. Salvador, *Phys. Rev.* **B39** (1989) 575.
[133] E. Manousakis and R. Salvador, *Phys. Rev. Lett.* **62** (1989) 1310.
[134] D. C. Mattis, *The Theory of Magnetism* (Springer, 1981).
[135] N. D. Mermin, *Ann. Phys. (N.Y.)* **21** (1963) 99.
[136] N. D. Mermin and H. Wagner, *Phys. Rev. Lett.* **17** (1966) 1133.
[137] G. Morandi, E. Galleani D'Agliano, F. Napoli and C. F. Ratto, *Adv. Phys.* **23** (1974) 867.
[138] N. F. Mott, *Proc. Phys. Soc.* **A62** (1949) 416.
[139] N. F. Mott, *Adv. Math. Phys.* **3** (1952) 76.
[140] N. F. Mott, *Canad. J. Phys.* **34** (1956) 1356.
[141] L. Neel, *Ann. Phys. (Paris)* **17** (1932) 64.
[142] L. Neel, *J. Phys. Radium* **3** (1932) 160.
[143] H. B. Nielsen and M. Ninomiya, *Nucl. Phys.* **B185** (1981) 20.
[144] H. B. Nielsen and M. Ninomiya, *Nucl. Phys.* **B193** (1981) 173.
[145] S. Okubo, *Phys. Rev.* **D16** (1977) 3535.
[146] W Pauli, *Nuovo Cim.* **6** (1957) 204.
[147] A. M. Polyakov, *Mod. Phys. Lett.* **A3** (1988) 325.
[148] A. P. Polychronakos, Florida University Preprint UFIFT-89-7, 1989.
[149] H. Primas, *Helv. Phys. Acta* **34** (1961) 331.
[150] H. Primas, *Rev. Mod. Phys.* **35** (1963) 710.
[151] R. Rajaraman, *Solitons and Instantons*, (North-Holland, 1982).
[152] A. N. Redlich, *Phys. Rev. Lett.* **52** (1984) 18.
[153] A. N. Redlich, *Phys. Rev.* **D29** (1984) 2366.
[154] J. D. Reger and A. P. Young, *Phys. Rev.* **B37** (1988) 5978.
[155] J. Schonfeld, *Nucl. Phys.* **B185** (1981) 157.
[156] J. R. Schrieffer, *Theory of Superconductivity*, (Benjamin, 1964).
[157] J. R. Schrieffer and P. A. Wolff, *Phys. Rev.* **149** (1966) 491.
[158] A. Schwartz, *Lett. Math. Phys.* **2** (1978) 247.
[159] G. F. Semenoff, *Phys. Rev. Lett.* **61** (1988) 517.
[160] M. A. Semenoff, Tjan-Sankii and L. D. Faddeev, *Vestnick Leningrad Univ. Math.* **10** (1982) 319.
[161] W. Siegel, *Nucl. Phys.* **B156** (1979) 135.
[162] R. R. P. Singh, M. P. Gel'Fand and D. A. Huse, *Phys. Rev. Lett.* **61** (1988) 2484.
[163] L. Smolin, *Mod. Phys. Lett.* **A4** (1989) 1091.
[164] L. Smolin, in *Knots, Topology and Quantum Field Theory*, Proc. of 13th Johns Hopkins Workshop, (World Scientific, 1990).
[165] J. Sokoloff, *Ann. Phys. (N.Y.)* **45** (1967) 186.
[166] J. Spalek, *Phys. Rev.* **B37** (1988) 533.
[167] N. Steenrod, *The Topology of Fibre Bundles* (Princeton University Press, 1951). See also A. P. Balachandran, G. Marmo, B. S. Skagerstam and A. Stern, *Gauge Symmetries and Fibre Bundles, Applications to Particle Dynamics*, (Springer, 1983).
[168] A. Stern, *Phys. Rev. Lett.* **59** (1987) 1506.
[169] A. Stern, Tuscaloosa Preprint UAHEP-884, 1988.
[170] R. L. Stratonovich, *Sov. Phys. Doklady* **2** (1958) 416.
[172] B. Sutherland, *An Introduction to the Bethe Ansatz*, unpublished Lecture Notes, 1988.
[172] D. Vollhardt, *Rev. Mod. Phys.* **56** (1984) 99.
[173] G. H. Wannier, *Phys. Rev.* **52** (1937) 191.

[174] S. Weinberg, *Prog. Theor. Phys. (Suppl.)* **86** (1986) 43.
[175] X. G. Wen and A. Zee, *Phys. Rev. Lett.* **61** (1988) 1025.
[176] X. G. Wen and A. Zee, *Phys. Rev. Lett.* **62** (1989) 2873.
[177] X. G. Wen and A. Zee, Santa Barbara Preprint NSF-ITP-88-145, 1988.
[178] X. G. Wen and A. Zee, Santa Barbara Preprint NSF-ITP-89-47, 1989.
[179] X. G. Wen, F. Wilczek and A. Zee, *Phys. Rev.* **B39** (1989) 11413.
[180] J. Wess and B. Zumino, *Phys. Lett.* **B37** (1971) 986.
[181] P. B. Wiegmann, *Phys. Lett.* **B60** (1988) 821.
[182] F. Wilczek, *Phys. Rev. Lett.* **48** (1982) 1144.
[183] F. Wilczek, *Phys. Rev. Lett.* **49** (1982) 957.
[184] F. Wilczek, "Fractional Statistics and Anyon Superconductivity", in *Anomalies, Phases, Defects . . .*", eds. M. Bregola, G. Marmo and G. Morandi, (Bibliopolis, Naples, 1990).
[185] F. Wilczek and A. Zee, *Phys. Rev. Lett.* **51** (1983) 2250.
[186] E. Witten, *Nucl. Phys.* **B311** (1988) 46.
[187] E. Witten, *Nucl. Phys.* **B323** (1989) 113.
[188] K. Wu, L. Yu and C. J. Zhu, *Mod. Phys. Lett.* **B2** (1988) 979.
[189] Y. S. Wu and A. Zee, *Phys. Lett.* **B147** (1984) 325.
[190] C. N. Yang, *Rev. Mod. Phys.* **34** (1962) 694.
[191] G. Zemba, MIT Preprint CTP-1721, 1989.
[192] J. Ziman, *Principles of the Theory of Solids* (Cambridge University Press, 1969).
[193] Z. Zou, *Phys. Lett.* **A131** (1988) 197.
[194] Z. Zou, Stanford Preprint, 1988.
[195] Z. Zou, *Phys. Rev.* **B40** (1989) 2262.
[196] Z. Zou, B. Doucot and B. S. Shastry, Princeton Preprint, 1988.
[197] B. Zumino, in *Relativity, Groups and Topology II*, eds. B. S. DeWitt and R. Stora (North-Holland, 1984).

Appendix A. BCS and Projected-BCS States

As is well known [156], a BCS state can be written in the form:

$$|\psi\rangle = \prod_{\mathbf{k}} (u_{\mathbf{k}} + v_{\mathbf{k}} b_{\mathbf{k}}^{\dagger})|0\rangle \tag{A.1}$$

where:

$$|u_{\mathbf{k}}|^2 + |v_{\mathbf{k}}|^2 = 1 \quad \forall \mathbf{k} \tag{A.2}$$

and the u's and v's are related to the gap parameter by:

$$\bar{u}_{\mathbf{k}} v_{\mathbf{k}} = \frac{\Delta_{\mathbf{k}}}{2E_{\mathbf{k}}} ; \quad E_{\mathbf{k}} = [(\epsilon_{\mathbf{k}} - \mu)^2 + |\Delta_{\mathbf{k}}|^2]^{1/2} \tag{A.3}$$

the $\epsilon_{\mathbf{k}}$'s being the single-particle energies, and μ the chemical potential.

If the reduced BCS Hamiltonian (including the chemical potential) is written as:

$$\mathcal{H}_{\text{BCS}} = \sum_{\mathbf{k}\sigma} (\epsilon_{\mathbf{k}} - \mu) c_{\mathbf{k}\sigma}^{\dagger} c_{\mathbf{k}\sigma} + \sum_{\mathbf{k}\mathbf{k}'} V_{\mathbf{k}\mathbf{k}'} c_{\mathbf{k}\uparrow}^{\dagger} c_{-\mathbf{k}\downarrow}^{\dagger} c_{-\mathbf{k}\downarrow} c_{\mathbf{k}'\uparrow} \tag{A.4}$$

the gap equation is, at $T = 0$:

$$\Delta_{\mathbf{k}} = -\sum_{\mathbf{k}'} V_{\mathbf{k}\mathbf{k}'} \frac{\Delta_{\mathbf{k}'}}{2E_{\mathbf{k}'}} \tag{A.5}$$

and:

$$\bar{u}_{\mathbf{k}} v_{\mathbf{k}} = \langle \psi | c_{-\mathbf{k}\downarrow} c_{\mathbf{k}\uparrow} | \psi \rangle . \tag{A.6}$$

The gap is determined by Eq. (A.5) only up to an overall phase factor, and so is the relative phase of $u_{\mathbf{k}}$ and $v_{\mathbf{k}}$. Apart from the phase, Eq. (A.5) will be assumed to have (and usually has) a unique solution.

It is a simple matter to show that the phase ambiguity is a direct consequence of the broken U(1) invariance which characterizes superconductivity. Indeed, the Hamiltonian (A.4) (as well as the full many-body Hamiltonian it originates from) is invariant under the global U(1) transformation:

$$c_{\mathbf{k}\sigma}^{\dagger} \to \exp[i\phi] \cdot c_{\mathbf{k}\sigma}^{\dagger} \tag{A.7}$$

with ϕ an angle defined mod (2π). This induces, via Eqs. (A.3) and (A.6), the phase change:

$$\Delta_{\mathbf{k}} \to \exp[-2i\phi] \cdot \Delta_{\mathbf{k}} \tag{A.8}$$

and defines a new BCS ground state:

$$|\psi\rangle \rightarrow |\psi; \phi\rangle =: \prod_{\mathbf{k}} [u_{\mathbf{k}} + v_{\mathbf{k}} e^{2i\phi} b_{\mathbf{k}}^{\dagger}]|0\rangle . \tag{A.9}$$

Equation (A.9) defines the manifold of the degenerate BCS ground states corresponding to the solutions of the self-consistency Eq. (A.5). The appearance of the factor of 2 in the additional phase reflects the fact that the global U(1) symmetry is actually broken down to $\mathbb{Z}_2$ in the superconducting state [174].

After these preliminaries, we now discuss the structure of the projections of the BCS state onto the subspaces containing a fixed number of particles (necessarily an even one). Equation (A.1) can be written as:

$$|\psi\rangle = A \prod_{\mathbf{k}} [1 + g(\mathbf{k}) b_{\mathbf{k}}^{\dagger}]|0\rangle \tag{A.10}$$

where:

$$g(\mathbf{k}) =: \frac{v_{\mathbf{k}}}{u_{\mathbf{k}}} ; \quad A =: \prod_{\mathbf{k}} u_{\mathbf{k}} . \tag{A.11}$$

Note that [156] $|u_{\mathbf{k}}|$ goes rapidly to one for $|\mathbf{k}| \gg k_F$ (with k_F the Fermi momentum), and hence the infinite product defining the normalization factor A is indeed well defined. The fact that $(b_{\mathbf{k}}^{\dagger})^2 = 0$ allows us to write:

$$1 + g(\mathbf{k}) b_{\mathbf{k}}^{\dagger} \equiv \exp[g(\mathbf{k}) b_{\mathbf{k}}^{\dagger}] \tag{A.12}$$

and hence:

$$|\psi\rangle = A \cdot \exp\left[\sum_{\mathbf{k}} g(\mathbf{k}) b_{\mathbf{k}}^{\dagger}\right]|0\rangle . \tag{A.13}$$

Therefore:

$$|\psi\rangle = A \sum_{n=0}^{\infty} \frac{1}{n!} \left(\sum_{\mathbf{k}} g(\mathbf{k}) b_{\mathbf{k}}^{\dagger}\right)^n |0\rangle . \tag{A.14}$$

Again, because of $(b_{\mathbf{k}}^{\dagger})^2 = 0$, we can rewrite:

$$\left[\sum_{\mathbf{k}} g(\mathbf{k}) b_{\mathbf{k}}^{\dagger}\right]^n \equiv \sum_{(\mathbf{k})}{}' g(\mathbf{k}_1) \ldots g(\mathbf{k}_n) b_{\mathbf{k}_1}^{\dagger} \ldots b_{\mathbf{k}_n}^{\dagger} \tag{A.15}$$

where $\sum_{\mathbf{k}}'$ stands for a multiple sum over $\mathbf{k}_1 \ldots \mathbf{k}_n$, with $\mathbf{k}_i \neq \mathbf{k}_j \ \forall i \neq j$.

The nth order term in the r.h.s. of (A.14) corresponds clearly to the component of $|\psi\rangle$ in the subspace of the Fock space containing $2n$ electrons. Calling P_{2n} the corresponding projection, we have then:

$$P_{2n}|\psi\rangle = \frac{A}{n!} \sum_{(\mathbf{k})}' g(\mathbf{k}_1) \ldots g(\mathbf{k}_n) b_{\mathbf{k}_1}^\dagger \ldots b_{\mathbf{k}_n}^\dagger |0\rangle \qquad (A.16)$$

(the factor of $(n!)^{-1}$ on the r.h.s. takes care of the fact that, for any given set $\{k_1 \ldots k_n\}$ of momenta, the summand is invariant under the $n!$ permutations of the factors). The norm of (A.16) is given by:

$$\|P_{2n}|\psi\rangle\|^2 = \frac{|A|^2}{n!} \sum_{(\mathbf{k})}' |g(\mathbf{k}_1)|^2 \ldots |g(\mathbf{k}_n)|^2 \qquad (A.17)$$

Defining then the normalized projection as:

$$|\psi_{2n}\rangle =: \frac{P_{2n}|\psi\rangle}{\|P_{2n}|\psi\rangle\|} \qquad (A.18)$$

we find:

$$|\psi_{2n}\rangle = \frac{A}{|A|} \left[\sum_{(\mathbf{k})} |g(\mathbf{k}_1)|^2 \ldots |g(\mathbf{k}_n)|^2 \right]^{-1/2}$$

$$\times \frac{1}{\sqrt{n!}} \sum_{(\mathbf{k})}' g(\mathbf{k}_1) \ldots g(\mathbf{k}_n) b_{\mathbf{k}_1}^\dagger \ldots b_{\mathbf{k}_n}^\dagger |0\rangle . \quad (A.19)$$

Thus

$$|\psi\rangle = \sum_n A_{2n} |\psi_{2n}\rangle ; \qquad \langle \psi_{2n} | \psi_{2m} \rangle = \delta_{nm} \qquad (A.20)$$

where the amplitudes A_{2n} are given by:

$$A_{2n} = \frac{|A|}{\sqrt{n!}} \left[\sum_{(\mathbf{k})}' |g(\mathbf{k}_1)|^2 \ldots |g(\mathbf{k}_n)|^2 \right]^{1/2} . \qquad (A.21)$$

It is easy to check that:

$$\sum_{n=0}^\infty |A_{2n}|^2 = |A|^2 \prod_\mathbf{k} (1 + |g(\mathbf{k})|^2) = 1 \qquad (A.22)$$

as it should be.

A given choice of phases (fixing, e.g., $\Delta_\mathbf{k}$ to be real for some fiducial value of $\mathbf{k}$) fixes the u's and v's uniquely, and hence both the reference state $|\psi\rangle$ of Eq. (A.1) and the $|\psi_{2n}\rangle$'s. It is easy to convince oneself that:

$$|\psi; \phi\rangle = \sum_{n=0}^\infty A_{2n} e^{2in\phi} |\psi_{2n}\rangle \equiv \prod_\mathbf{k} (u_\mathbf{k} + v_\mathbf{k} e^{2i\phi} b_\mathbf{k}^\dagger) |0\rangle . \qquad (A.23)$$

We have then the important results [156]:

$$P_{2n}|\psi\rangle = A_{2n}|\psi_{2n}\rangle = \int_0^{2\pi} \frac{d\phi}{2\pi} e^{-2in\phi}|\psi;\phi\rangle \qquad (A.24)$$

and

$$|A_{2n}|^2 = \int_0^{2\pi} \frac{d\phi}{2\pi} e^{-2in\phi}\langle\psi|\psi;\phi\rangle \equiv \int_0^{2\pi} \frac{d\phi}{2\pi} e^{-2in\phi} \prod_{\mathbf{k}} (|u_{\mathbf{k}}|^2 + |v_{\mathbf{k}}|^2 e^{2i\phi}) . \qquad (A.25)$$

Note also that, from Eq. (A.23), we have

$$\langle\psi;\phi|\psi;\phi'\rangle = \sum_{n=0}^{\infty} |A_{2n}|^2 \exp[-2in(\phi - \phi')] . \qquad (A.26)$$

If $\bar{N} =: \langle\hat{N}\rangle$ is the average particle number, one can prove [43, 156] that the mean-square deviation $(\Delta N)^2 =: \langle[N - \bar{N}]^2\rangle$ grows as $\bar{N}$ for $\bar{N} \to \infty$. It can also be proved [43] that, in the same limit, $\langle\psi;\phi|\psi;\phi'\rangle \propto \exp[-4\delta\bar{N}(\phi - \phi')^2]$, where $\delta = (\Delta N)^2/\bar{N}$ remains finite for $\bar{N} \to \infty$. Therefore, the scalar product (A.26) vanishes in the thermodynamic limit for all $\phi' \neq \phi$. This is in agreement with the fact that, in the thermodynamic limit, the different (quasiparticle) vacua, labeled by the angle ϕ, correspond to different superselection sectors of the theory.

Again, under the U(1) transformation (A.7), $|\psi_{2n}\rangle$ changes only by a phase factor, namely:

$$|\psi_{2n}\rangle \to \exp[2in\phi] \cdot |\psi_{2n}\rangle . \qquad (A.27)$$

It follows from this that projection onto a subspace with fixed particle number restores the gauge invariance of the theory. No gauge-variant operator can have a nonzero expectation value between these states, and, in particular:

$$\langle\psi_{2n}|b_{\mathbf{k}}|\psi_{2n}\rangle = 0 \qquad (A.28)$$

and off-diagonal long-range order (ODLRO [68]) is destroyed by the projection. In view of Eq. (A.24), we can also say that ODLRO is destroyed by the process of taking averages over the phase of the gap parameter.

We now discuss briefly the nature of the excited states over the BCS ground state. Acting with the quasiparticle creation operator (4.18) on the ground state, we obtain, with some algebra [43, 156]:

$$\alpha_{\mathbf{k}}^\dagger|\psi\rangle = \prod_{\mathbf{q}\neq\mathbf{k}} (u_{\mathbf{q}} + v_{\mathbf{q}}b_{\mathbf{q}}^\dagger)c_{\mathbf{k}\uparrow}^\dagger |0\rangle \qquad (A.29)$$

$$\beta_{\mathbf{k}}^\dagger|\psi\rangle = \prod_{\mathbf{q}\neq\mathbf{k}} (n_{\mathbf{q}} + v_{\mathbf{q}}b_{\mathbf{q}}^\dagger)c_{-\mathbf{k}\downarrow}^\dagger |0\rangle \qquad (A.30)$$

while, e.g.:

$$\alpha_{\mathbf{k}}^\dagger \alpha_{\mathbf{k}'}^\dagger |\psi\rangle = \prod_{\mathbf{q}\neq \mathbf{k},\mathbf{k}'} (u_\mathbf{q} + v_\mathbf{q} b_\mathbf{q}^\dagger) c_{\mathbf{k}\uparrow}^\dagger c_{\mathbf{k}'\uparrow}^\dagger |0\rangle . \tag{A.31}$$

Similar results are obtained by apply other bilinears in the quasiparticle creation operators (see Ref. 43 for more details).

It is quite clear that states of the form (A.29)–(A.30) have an expansion in states with a number of particles of a parity opposite to that in the expansion of $|\psi\rangle$, while (A.31) (and, in general, states obtained by applying an even number of quasiparticle creation operators) will have an expansion in states of the same parity as $|\psi\rangle$. If we look then for the excited states over a projected (at fixed N) BCS state, they will be obtained by projecting onto the excited BCS states of the form (A.31) with N quasiparticles or, in general, states obtained by adding an even number of quasiparticles to the BCS ground state.

Appendix B. A Cursory Look at the Quantum Hall Effect

We shall consider here only electrons with charge e moving in a *uniform* and *constant* magnetic field **B** directed along the z-axis of a Cartesian reference frame: $\mathbf{B} = B\mathbf{k}$ ($|\mathbf{k}| = 1$). The electron dynamics along the z-axis is of course trivial and decouples from that in the orthogonal plane(s), so we shall consider only motions in the (x, y) plane. The Hamiltonian for free electrons in the presence of the field **B** can be written as:

$$\mathcal{H} = \frac{\Pi^2}{2m} \tag{B.1}$$

where:

$$\Pi =: -i\hbar\nabla + e\frac{\mathbf{A}}{c} \tag{B.2}$$

is the *kinetic* momentum ($\Pi = m \cdot \mathbf{v}$, and v is the quantum operator corresponding to the velocity), and **A** is the vector potential: $\mathbf{B} = \nabla \times \mathbf{A}$.

Working out the commutation relations, we find:

$$[\Pi_x, \Pi_y] = -i\hbar m\Omega \tag{B.3}$$

where: $\Omega =: eB/mc$ is the *cyclotron* (or *Larmor*) *frequency*. Defining next:

$$a =: \frac{1}{\sqrt{2m\hbar\Omega}}(\Pi_x - i\Pi_y) \tag{B.4}$$

one easily finds the commutation relation:

$$[a, a^\dagger] = 1 \tag{B.5}$$

and:

$$\mathcal{H} = \hbar\Omega\left[a^\dagger a + \frac{1}{2}\right]. \tag{B.6}$$

The spectrum of $\mathcal{H}$ is then that of a simple harmonic oscillator of proper frequency Ω, and the result is of course gauge independent.

The ground state manifold is spanned by the solutions of the equation:

$$a|0\rangle = 0 \; ; \quad \langle 0|0\rangle = 1 \tag{B.7}$$

and the (normalized) excited states are given by:

$$|n\rangle = (n!)^{-1/2}(a^\dagger)^n|0\rangle . \tag{B.8}$$

Of course, there will be a "tower" of excited states (labeled by the integer n) for each independent solution of Eq. (B.7).

We discuss now the detailed structure of the solutions in some relevant gauges, namely:

i) The *Landau* gauge, defined by:

$$\mathbf{A} = B(0, x, 0) \, . \tag{B.9}$$

In this gauge, $\hat{p}_y \equiv -i\hbar \nabla_y$ is a constant of the motion, and can be diagonalized along with $\mathcal{H}$. Adopting periodic boundary conditions in the y-direction (assumed to be of total length L), the eigenfunctions will be given by:

$$\Psi_{n,k}(x, y) = \frac{1}{\sqrt{L}} \exp[-iky]\phi_{n,k}(x) \, ; \qquad k = \frac{2\pi p}{L} \, ; \qquad p \in \mathbb{Z} \tag{B.10}$$

and the effective (one-dimensional) Hamiltonian acting on the reduced wavefunction $\phi_{n,k}(x)$ is:

$$\mathcal{H}_{\text{eff}} = -\frac{\hbar^2}{2m} \frac{d^2}{dx^2} + \frac{1}{2} m\Omega^2 (x - l^2 k)^2 \tag{B.11}$$

where l is the *magnetic length*:

$$l = \sqrt{\frac{\hbar c}{eB}} \tag{B.12}$$

leading to:

$$\phi_{n,k}(x) = N_n \exp\left[-\frac{(x - l^2 k)^2}{2l^2}\right] \cdot H_n\left(\frac{x}{l} - kl\right) \tag{B.13}$$

with N_n a normalization factor, and H_n the nth Hermite polynomial. For further use, let's remark that the magnetic length satisfies: $2\pi l^2 B = hc/e$, where the r.h.s. is a ratio of universal constants.

ii) The *symmetric gauge*, defined by:

$$\mathbf{A} =: \frac{1}{2} \mathbf{B} \times \mathbf{r} \, . \tag{B.14}$$

The Hamiltonian exhibits then cylindrical symmetry, and the z-component of the angular momentum:

$$\hat{L}_z = -i\hbar \left(x \cdot \frac{\partial}{\partial y} - y \cdot \frac{\partial}{\partial x} \right) \tag{B.15}$$

is conserved. Introducing complex variables $z, \bar{z}, z =: x + iy$, the ground-state equation becomes:

$$\left[\frac{\partial}{\partial \bar{z}} + \frac{z}{2l^2}\right] \Psi_0(z, \bar{z}) = 0 . \tag{B.16}$$

Setting then:

$$\Psi_0 =: \exp\left[-\frac{|z|^2}{4l^2}\right] \cdot \Phi_0 . \tag{B.17}$$

Equation (B.16) becomes:

$$\frac{\partial}{\partial \bar{z}} \Phi_0 = 0 . \tag{B.18}$$

Therefore, as far as the ground-state manifold is concerned, Φ_0 *can be a function of z alone*.

The simultaneous ground-state eigenfunctions of $\mathcal{H}$ and L_z are:

$$\Psi_{m,0}(x, y) = [(2l)^{m+1}\sqrt{\pi m!}]^{-1} z^m \exp\left[-\frac{\pi|z|^2}{4l^2}\right] . \tag{B.19}$$

Counting of States (In the Landau gauge, but the same results obtain in the symmetric gauge).

The wave vector k is quantized according to:

$$k = \frac{2\pi p}{L} ; \quad p \in \mathbb{Z} . \tag{B.20}$$

States of the form (B.13) are Gaussians centered around: $x_n =: l^2 k$. If the sample extends, in the x direction, from $-W/2$ to $+W/2$ for some (finite) W, then:

$$-\frac{W}{2} < x_n < +\frac{W}{2} \tag{B.21}$$

and the total number of states for fixed n (i.e. the degeneracy of a Landau level) is given by:

$$N_B = \frac{LW}{2\pi l^2} \tag{B.22}$$

while the *density of states* (number of states per unit area) for the nth level is given by:

$$n_B = (2\pi l^2)^{-1} \equiv \frac{eB}{hc} . \tag{B.23}$$

Remarks

i) The *elementary fluxon* (magnetic flux quantum) is defined by:

$$\Phi_0 =: \frac{hc}{e} \tag{B.24}$$

Then:

$$N_B \equiv \frac{\Phi}{\Phi_0}; \quad \Phi =: BLW \tag{B.25}$$

i.e. N_B is the number of elementary fluxons threading the sample, and will be always assumed to be an integer.

ii) If the sample contains n electrons per unit surface, the *filling factor* (or "*filling fraction*") ν is defined by:

$$\nu =: \frac{n}{n_B} \equiv \frac{nhc}{eB}. \tag{B.26}$$

Integer value of ν correspond then to (one or more) completely filled Landau levels. Note that one can also write the total number of electrons N as $N = \nu\Phi/\Phi_0$.

After these preliminaries, we now turn to the Hall effect, starting with a brief account of its conventional (classical) theory. Consider a slab of conducting material in crossed (static) electric and magnetic fields **E** and **B** (see Fig. 13). Let us also assume for simplicity the absence of damping mechanisms for the electron motion. Then, if $|\mathbf{E}| < |\mathbf{B}|$, there is a *net* current flow at right angles with both **E** and **B**, whose direction depends on the sign of the charges involved. If $\mathbf{E} = 0$, we know what happens from elementary textbooks: the electrons move in circular orbits whose radius is the "*Larmor radius*" (v/Ω) ($v = |\mathbf{v}|$ is a constant of the motion). The effects of a nonvanishing **E** can be "boosted away" by performing a Lorentz boost along the y-axis (see Fig. 13) with boost velocity:

$$v_D = \frac{cE}{B} \tag{B.27}$$

(as $E \ll B$ is typical experiments, actually $v_D \ll c$). The resulting trajectories are called "trochoids", and are superpositions of circular and uniform drift motions. The net current is therefore (with n the (2D) carrier density):

$$J_H = nv_De \equiv \frac{nec}{B} E \tag{B.28}$$

thus defining the *Hall conductivity*:

$$\sigma_H =: \frac{nec}{B}. \tag{B.29}$$

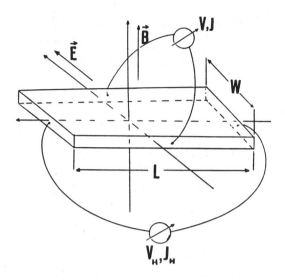

Fig. 13. A theorist's view of an experiment on the Hall effect: **E**, **B** = applied electric and magnetic fields; W, L = sample sizes; V, J = "normal" parallel potential drop and current; V_H, J_H = Hall voltage and current.

In two space dimensions, the ratio: e^2/h has the dimensions of a conductivity, and: $nech/Be^2 = \nu$, the filling factor. So, we obtain:

$$\frac{h}{e^2} \cdot \sigma_H = \nu . \quad (B.30)$$

Therefore, in "natural" units of e^2/h, $\sigma_H = \nu$, corresponding to the dotted line of Fig. 14. The by now famous results of von Klitzing, Dorda and Pepper are instead schematically represented by the full line of Fig. 14. Their main features are:
  i) Well defined *plateaux* (at low temperatures) centered around *integer* fillings, and
  ii) *Integer* values (in units of e^2/h) of the plateau Hall conductivity:

$$\sigma_H = M \frac{e^2}{h}; \quad M = 1, 2, \ldots . \quad (B.31)$$

The systems in which this *Integral Quantum Hall Effect* (IQHE) shows up (at high field and low temperatures) are:
  i) Essentially 2D systems (MOSFET's and heterostructures), and also, especially
  ii) Disordered systems (this being due to the preparation procedures).

We inquire now briefly on the effect of impurities and disorder on the structure of Landau levels. In a perfectly pure sample, and for **E** = 0, the density of states is composed of a series of δ functions centered at the position of the Landau levels $E_n = (n + 1/2)\hbar\Omega$, as shown in Fig. 15. It can be easily shown that the introduction of the electric field completely lifts the degeneracy of the Landau levels, modifying the spectrum from E_n to:

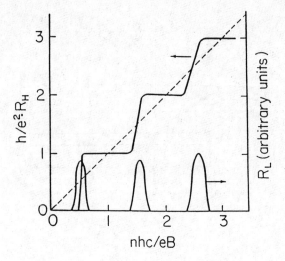

Fig. 14. A schematic (theorist's) view of the Integer Quantum Hall Effect.

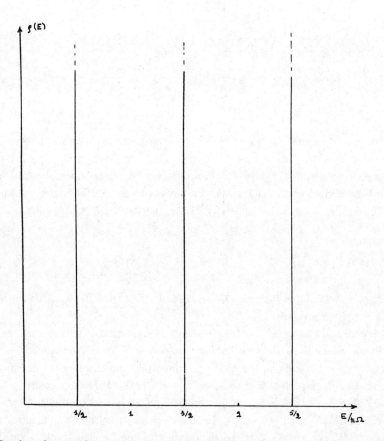

Fig. 15. Density of states for pure Landau levels. Vertical straight lines stand for δ-functions at $E = $ (half-integer) $\hbar\Omega$.

$$E_{n,k} = \left(n + \frac{1}{2}\right)\hbar\Omega + eEl^2k . \quad (B.32)$$

The original sharp Landau levels are then broadened into bands centered at $(n + 1/2)\hbar\Omega$ and of total width eEW, corresponding to the potential drop across the sample. As, under typical experimental conditions (that is for high magnetic fields), $eEW \ll \hbar\Omega$, the bands are well separated from each other (the separation remaining roughly of the order of $\hbar\Omega$).

The effect of impurities and disorder turns out to be that of splitting *localized states* off each band (from both above and below), which tend to fill in the gap between adjacent subbands, until the (qualitative) picture of the density of states looks as in Fig. 16. As long as the Fermi level lies within a tail of localized states, there are no electron states available to contribute to the conduction besides those which already do, and hence one should expect σ_H = const. under such conditions. On the contrary, the conductivity is expected to rise sharply when the Fermi level sweeps a new set of extended states. *The above picture for the density of states explains the plateau, but of course not the* **integer** *values of* $\sigma_H \cdot (h/e^2)$.

The main contributions towards a complete understanding of the integral quantization of the conductivity in the IQHE have been given by Thouless and coworkers, and by Avron, Seiler and Simon. We omit details here, but only quote the main results of the above authors.

By evaluating the conductivity via Linear Response Theory and the Kubo formula, Thouless and coworkers have proved a result which turns out to be crucial for all subsequent treatments, namely that *the Hall conductivity* $\sigma_H(\mathbf{r})$ *at any point* $\mathbf{r}$ *inside the sample is a* **local** *quantity, depending only on the properties of the medium in an*

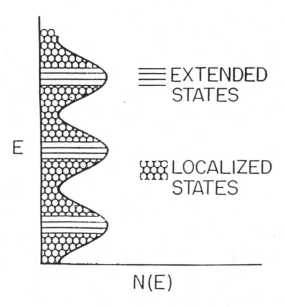

Fig. 16. The actual density of states in a disordered sample, showing tails of localized states.

immediate neighborhood of **r**, *with corrections dying away exponentially on the scale of the magnetic length*. The main consequence of this result is that the Hall conductivity is *essentially insensitive to the boundary conditions* one imposes at the edges of the physical samples. The Hall conductivity one measures in actual experiments is given by the sample-averged conductivity:

$$\sigma_H = \frac{1}{A} \int d^2 r \sigma_H(\mathbf{r}) \; ; \quad A = LW \; . \tag{B.33}$$

Under the assumption that *the Fermi level lies in a tail of localized states*, Thouless and coworkers found, for the sample-averaged Hall conductivity:

$$\sigma_H = \sum_n f(\epsilon_n) \sigma_{H,n} \tag{B.34}$$

where the ϵ_n's are the single-particle energy eigenvalues, n is a Landau-level index, $f(.)$ is the Fermi function, and:

$$\sigma_{H,n} = \frac{i\hbar e^2}{A} \sum_{m \neq n} \frac{(v_x)_{nm}(v_y)_{mn} - (v_y)_{nm}(v_x)_{mn}}{(\epsilon_n - \epsilon_m)^2} \; . \tag{B.35}$$

Recall that the velocity operator **v** is given by:

$$\mathbf{v} = \frac{1}{m} \left[\frac{\hbar}{i} \nabla - \frac{e\mathbf{A}}{c} \right] \; . \tag{B.36}$$

Adding a pure gauge field, i.e. letting

$$\nabla \to \nabla + \frac{i}{\hbar} \boldsymbol{\lambda} \; ; \quad \boldsymbol{\lambda} = \text{const} \; , \tag{B.37}$$

the gauge field can be reabsorbed by redefining the boundary conditions as:

$$\Psi(x + W) = \exp\left[-i \frac{\lambda_x W}{\hbar}\right] \Psi(x) \; ,$$
$$\Psi(y + L) = \exp\left[-i \frac{\lambda_y L}{\hbar}\right] \Psi(y) \; . \tag{B.38}$$

The wavefunctions depend then *periodically* (with period 2π) on the angles:

$$\alpha =: \frac{\lambda_x W}{\hbar} \; , \quad \beta =: \frac{\lambda_y L}{\hbar} \; . \tag{B.39}$$

Moreover:

$$v_x = \frac{\partial \mathcal{H}}{\partial \lambda_x} = \frac{W}{\hbar} \frac{\partial \mathcal{H}}{\partial \alpha} \; ; \qquad v_y = \frac{\partial \mathcal{H}}{\partial \lambda_y} = \frac{L}{\hbar} \frac{\partial \mathcal{H}}{\partial \beta} \; . \qquad (B.40)$$

We obtain therefore:

$$\sigma_{H,n} = 2\pi i \frac{e^2}{h} \sum_{m \neq n} \frac{1}{(\epsilon_m - \epsilon_n)^2} \left\{ \left\langle n \left| \frac{\partial \mathcal{H}}{\partial \alpha} \right| m \right\rangle \left\langle m \left| \frac{\partial \mathcal{H}}{\partial \beta} \right| n \right\rangle - (\alpha \Leftrightarrow \beta) \right\} . \qquad (B.41)$$

Avron, Seiler and Simon have further shown that (B.41) can be recast into the more compact form:

$$\sigma_{H,n} = 2\pi i \frac{e^2}{h} \left\{ \left\langle \frac{\partial n}{\partial \alpha} \Big| \frac{\partial n}{\partial \beta} \right\rangle - \left\langle \frac{\partial n}{\partial \beta} \Big| \frac{\partial n}{\partial \alpha} \right\rangle \right\} . \qquad (B.42)$$

In view of the stated independence of the Hall conductivity on the boundary conditions, one can replace (B.42) by its average on the boundary conditions, thereby obtaining eventually:

$$\sigma_{H,n} = \frac{e^2}{h} \int_0^{2\pi} \int_0^{2\pi} \frac{\Omega_n}{2\pi i} \qquad (B.43)$$

where the two-form Ω_n is given by:

$$\Omega_n = dA_n \; ; \qquad A_n = \langle dn | n \rangle \; . \qquad (B.44)$$

In this form, Ω_n is easily recognized to be *the curvature two-form of a connection A_n on a principal* U(1) *bundle over the base space parametrized by the angles α and β.* $(h/e^2) \cdot \sigma_{H,n}$ is then *the first Chern number associated with the above connection, and is hence an integer.* As, at low temperatures: $f(\epsilon) \simeq \theta(\epsilon_F - \epsilon)$, ϵ_F being the Fermi energy, the total Hall conductivity will be an integer as well.

From the form of the connection, one can further conclude that $(h/e^2) \cdot \sigma_{H,n}$ can also be viewed as the Berry phase associated with an adiabatic circuit along the boundary ($\alpha = 0$ or 2π and/or $\beta = 0$ or 2π) of the set of boundary conditions labeled by the angles α and β.

We now turn to a brief account of the *Fractional* Quantum Hall Effect (FQHE). This effect was discovered about two years after the IQHE. It occurs in highly pure samples, at lower temperatures (~ 1 K) and higher fields (~ 15 Teslas) than the IQHE, again with plateau and exact quantization of the Hall conductivity, namely:

$$\sigma_H = \nu \frac{e^2}{h} \qquad (B.45)$$

but with $\nu = p/q$, p and q being relatively prime integers. Further, apart from some recent claims, q is an *odd* integer, with typical values $\nu = 1/3$ (the best known experimental value), 2/5, 2/7 etc. We shall concentrate here on the (by now) "classic" $\nu = 1/3$ case.

The status of the topological analysis of the effect is definitely not as clear as it is for the IQHE, although some authors have argued that, *if* there is a gap of extended states at $\nu = p/q$, then the ground state should be q-fold degenerate, and that an extension of the topological analysis should lead to fractional quantization.

It seems however that the most widely accepted "paradigm" for the structure of the ground state, due to Laughlin, leaves very little elbow room for degeneracy, and therefore that the explanation for the experimentally found fractional quantization calls for some kind of entirely different theoretical approach.

We shall now discuss briefly Laughlin's approach to the description of the ground state and of the (low lying) excited states appropriate for the FQHE. Laughlin's ground state wavefunction is a variational function, meant to describe an assembly of *strongly correlated* electrons in the lowest ($n = 0$) Landau level. The description of the electron system departs then considerably from that appropriate to the IQHE which, we recall, is in terms of essentially independent (quasi) particles subject to impurity scattering. Here it is believed that Coulomb correlations play a dominant role, and that *gaps at rational fillings are entirely due to many body effects*.

With reference to what has already been discussed, the single particle wavefunctions for the lowest Landau level, and in the symmetric gauge, can be written as:

$$\psi_{p,0}(z) = \tilde{N}_{p,0} \exp\left[\frac{|z|^2}{4l^2}\right] (\partial_{\bar{z}})^p \exp\left[-\frac{|z|^2}{2l^2}\right] \tag{B.46}$$

with $\tilde{N}_{p,0}$ a normalization factor. If the lowest Landau level is entirely filled ($\nu = 1$), the ground state wavefunction is given, in the absence of correlations, by the Slater determinant of the single particle wavefunctions (B.46) for $p = 0, \ldots, N_B - 1$. The latter turns out to be essentially a Vandermonde determinant, and we have, calling Ψ_0 the ground-state wavefunction, and N^* a normalization factor:

$$\Psi_0 = N^* \prod_{i<j} (z_i - z_j) \exp\left[-\sum_{k=1}^{N} \frac{|z_k|^2}{4l^2}\right]. \tag{B.47}$$

Note that the (z-component of) the angular momentum operator can be written as:

$$\hat{L} = \hbar \sum_{i=1}^{N} (z_i \partial_{z_i} - \bar{z}_i \partial_{\bar{z}_i}) \tag{B.48}$$

so that:

$$\hat{L}\Psi_0 = \frac{N(N-1)}{2} \hbar \Psi_0 \tag{B.49}$$

that is, Ψ_0 is an eigenfunction of the total angular momentum.

In the first of many seminal papers, Laughlin proposed a trial wavefunction of the form:

$$\Psi = N^* \prod_{i<j} f(z_i - z_j) \exp\left[-\sum_k \frac{|z_k|^2}{4l^2}\right] \qquad (B.50)$$

where the *a priori* unknown function f should satisfy the following requirements:

i) Having to describe electrons confined to the lowest Landau level, f should be *a function of the z's alone* (and not of the $\bar{z}$'s);

ii) As a consequence of Pauli's principle, f should be *odd under any interchange* $i \Leftrightarrow j$, and

iii) f should be also *an eigenfunction of the total angular momentum* (which commutes with the total Hamiltonian in the case of interactions such as the Coulomb interction as well).

All this boils down to f being of the form:

$$f(\zeta) = \zeta^m, \qquad m \text{ an } odd \text{ integer} \qquad (B.51)$$

and we end up with the final form of Laughlin's wavefunction, namely:

$$\Psi = N^* \prod_{i<j} \left[\frac{z_i - z_j}{l}\right]^m \exp\left[-\sum_k \frac{|z_k|^2}{4l^2}\right]. \qquad (B.52)$$

The wavefunction Ψ of Eq. (B.52) is now an eigenfunction of the total angular momentum with eigenvalue $N(N-1)m\hbar/2$.

The probability density can be written as:

$$|\Psi|^2 = \text{const.} \cdot \exp[-\beta\Phi] \qquad (B.53)$$

where: $\beta = 1/m$, and:

$$\Phi =: -2m^2 \sum_{i<j} \ln\left|\frac{z_i - z_j}{l}\right| + \frac{1}{2} m \sum_i \frac{|z_i|^2}{l^2}. \qquad (B.54)$$

In this form, $|\Psi|^2$ can be viewed as the *classical probability distribution of a one-component plasma (OCP)* at "temperature" $\beta = 1/m$, the plasma being made up of particles of charge: $Q = m$ which:

i) Repel each other via logarithmic interactions (the natural form of the Coulomb interaction in two space dimensions) and

ii) interact with a fixed background of charge of opposite sign and charge density:

$$\sigma = (2\pi l^2)^{-1} \equiv n_B. \qquad (B.55)$$

The value of m is then fixed by the requirement of *electrical neutrality* of the system. If there are n particles per unit surface area, then we must have: $n \cdot m = n_B$, and, as $\nu = (n/n_B)$, we eventually find:

$$\nu = \frac{1}{m} \tag{B.56}$$

establishing the link between the filling fraction and the exponent m.

Elementary excitations in the nature of quasiparticles or quasiholes can be created in the following manner. Consider again the free-electron system, with eigenfunctions (B.46). It can be proved that, if one pierces the plane at $z = 0$ with an "infinitely thin" solenoid, and varies adiabatically the flux inside the solenoid from zero to $\Phi_0 = hc/e$ (that is by one flux quantum), the result is an adiabatic map of state p into state $(p + 1)$ $(p - 1)$ if the flux is varied by Φ_0, with the state labeled by $p = 0$ being mapped into the next Landau level). If one now removes the solenoid (remember that, if Φ/Φ_0 = integer, this can be done with the aid of a nonsingular gauge transformation), the net result of this "Gedankenexperiment" will be to leave a hole at the place were the solenoid was (a particle if $\Phi = -\Phi_0$). By switching on the interaction adiabatically, the state will evolve into an excited state of the *full* Hamiltonian containing just one quasihole (this is in the same spirit as the description of quasiholes (or quasiparticles) in Landau's Fermi Liquid theory). Generalizing (trivially) this argument to a quasihole centered at an arbitrary point z_0, Laughlin has proposed the (approximate) wavefunction for one hole at z_0 in the form:

$$\Psi_{z_0}^+ = N_+ A_{z_0} \Psi \; ; \qquad A_{z_0} =: \prod_{i=1}^{N} (z_i - z_0) \; . \tag{B.57}$$

The adjoint operator:

$$A_{z_0}^\dagger =: \prod_{i=1}^{N} \left(\frac{\partial}{\partial z_i} - \frac{\bar{z}_0}{l^2} \right) \tag{B.58}$$

will in turn create a quasiparticle at z_0. Note that A_{z_0} and $A_{z'_0}^\dagger$ do *not* commute for $z_0 \neq z'_0$. This is connected with the fact that the quasiparticles have finite size, and hence the processes of creating or destroying them at different points are not independent processes.

It has been proved that:

i) Comparison with numerical calculations gives excellent agreement between (B.57) and the exact wavefunctions (calculated for systems with a finite number of (actually few) particles, of course).

ii) One can actually construct model Hamiltonians for which both Laughlin's ground state and excited states are *exact* eigenstates.

We want to discuss now both the charge and statistics of Laughlin's quasiparticles [26].

One way of measuring the charge of a particle is to carry it adiabatically around a solenoid enclosing a total flux Φ. Then, if we denote by e^* the particle's charge, the *adiabatic (Berry) phase* its wavefunction will acquire is given by:

$$\Delta\gamma = \frac{e^*}{\hbar c}\int \mathbf{A}\cdot \mathbf{dl} \equiv 2\pi\frac{e^*}{e}\cdot\frac{\Phi}{\Phi_0} \qquad (B.59)$$

where $\mathbf{A}$ is the vector potential, and the integral is taken along the adiabatic circuit. The phase (B.59) can be measured in an Aharonov-Bohm-type interference experiment. On the other hand, if $|\psi(t)\rangle$ is the instantaneous wavefunction, Berry's phase can be calculated from the parallel-transpot equation:

$$\frac{d\gamma}{dt} = i\left\langle \psi(t)\left|\frac{d\psi}{dt}\right.\right\rangle. \qquad (B.60)$$

Consider now the process in which a quasihole located initially at z_0 is adiabatically dragged along a circle of radius $R = |z_0| \gg l$ centered at the origin. The adiabatic circuit will enclose a flux Φ determined, at filling ν, by the condition:

$$\langle n \rangle_R = \nu\frac{\Phi}{\Phi_0} \qquad (B.61)$$

where $\langle n \rangle_R$ is the average number of the electrons inside the disk of radius R. Employing the wavefunction (B.57), we have at once:

$$\frac{d}{dt}\Psi_{z_0}^+ = \sum_i \frac{d}{dt}\ln(z_i - z_0(t))\Psi_{z_0}^+ \qquad (B.62)$$

and hence:

$$\frac{d\gamma}{dt} = i\left\langle \Psi_{z_0}^+ \left| \frac{d}{dt}\sum_i \ln(z_i - z_0(t))\right| \Psi_{z_0}^+\right\rangle. \qquad (B.63)$$

Introducing the average electron density in state Ψ_{z_0} as:

$$\rho_+(z) = \left\langle \Psi_{z_0}^+ \left| \sum_i \delta(z - z_i)\right| \Psi_{z_0}^+\right\rangle \qquad (B.64)$$

we obtain:

$$\frac{d\gamma}{dt} = i\int dxdy\rho_+(z)\frac{d}{dt}\ln(z - z_0(t)). \qquad (B.65)$$

We expect $\rho_+(z)$ to be of the form:

$$\rho_+(z) = \rho_0 + \delta\rho_+(z) \qquad (B.66)$$

with: $\pi R^2 \rho_0 = \langle n \rangle_R$, and $\delta\rho_+$ a localized correction with spatial extent of the order of the size of the quasihole. The total variation of $\ln(z - z_0)$ will be $2\pi i$ if $|z| < R$, zero otherwise, and we obtain:

$$\Delta\gamma = -2\pi\langle n \rangle_R + \text{finite corrections} \qquad (B.67)$$

and, as $\langle n \rangle_R \propto R^2$:

$$\Delta\gamma = -2\pi\langle n \rangle_R = -2\pi\nu\frac{\Phi}{\Phi_0} \qquad (B.68)$$

in the limit of large R's. Comparing with Eq. (B.62), we see that the *quasiholes carry a fractional charge* given by:

$$e^* = -\nu e \ . \qquad (B.69)$$

The same will be true for the quasiparticles, with a charge of opposite sign and same magnitude.

Consider now two quasiholes located at z_a and z_b respectively, with $|z_a - z_b| = R$, and the process of dragging hole "b" adiabatically around hole "a". The fact that the hole at z_a carries a charge $-\nu e$ can be interpreted as implying that *exactly ν electrons* have been removed from the (big) disk of radius R centered at z_a. The same considerations as before apply, but for the fact that $\langle n \rangle_R$ must be substituted by $\langle n \rangle_R - \nu$, and *an extra phase (of statistical origin)*

$$\Delta\gamma' = 2\pi\nu \qquad (B.70)$$

is accumulated in the process. It should be clear from the previous discussion that the extra phase is accumulated steadily during the adiabatic motion. If we consider now the process of *exchange* of two quasiholes, a process which can be accomplished by letting each of them make a π turn around the other, *the total phase change will be $\pi\nu$*. For $\nu = 1 \pmod{2\pi}$ the quasiholes (as well as the quasiparticles) will be fermions, while they will be bosons for $\nu = 0$ (again mod 2π). For *fractional* fillings, however, they will obey *fractional (or "anyon") statistics*.

Elementary excitations above Laughlin's ground state are to be considered as identical particles moving in a 2D manifold (essentially $\mathbb{R}^2$). Their (inequivalent) scalar quantizations are associated with the one-dimensional unitary representations of the corresponding braid group [45] ($\mathbb{B}_n$ for n particles), which are known to be classified by an angle θ, $0 < \theta < 2\pi$. Not only does the example of the FQHE show a concrete physical realization of "braid statistics", it shows also that, among all the "kinematically" equally possible values of θ, the actual dynamics (and the stability of the ground state) acts to select a specific value of θ, namely: $\theta = \pi\nu$, where $1/\nu$ is an odd integer.

The literature on both the Integer and the Fractional Quantum Hall Effects is so vast that we cannot quote all of it here. We refer the interested reader only to the comprehensive and up-to-date review of all the aspects that have been discussed here contained in: "The Quantum Hall Effect" 2nd edition, eds. R. E. Prange and S. M. Girvin, (Springer-Verlag, 1990) (see also: G. Morandi: "Quantum Hall Effect", (Bibliopolis, 1988)).

Appendix C. A Continuum Model for a System of Free Spins

It seems appropriate here to briefly mention certain developments in the investigations of a system of free spins in a plane.

A novel continuum chiral model in 2+1 dimensions was constructed in Ref. [34] which could be associated with a system of free spins in a plane. The Lagrangian considered in Ref. [34] (see also Eq. (8.29)) for a lattice of noninteracting spins located at positions $\mathbf{x}_i$ was taken as

$$\mathscr{L} = \int d^2x \rho_l(\mathbf{x}) ij \, \text{Tr}[\sigma_3 g(\mathbf{x}, t)^{-1} \dot{g}(\mathbf{x}, t)] \qquad (C.1)$$

where $g(\mathbf{x}, t) \in SU(2)$, σ_α stand for Pauli matrices and

$$\rho_l(\mathbf{x}) = \sum_i \delta^{(2)}(\mathbf{x} - \mathbf{x}_i) \qquad (C.2)$$

$\rho_l(\mathbf{x})$ can be interpreted as the number density of spins at $\mathbf{x}$, and the subscript l denotes that this number density corresponds to the lattice model. A continuum model associated with this spin lattice can be constructed if one can find a smooth density function ρ to replace the discretely supported ρ_l, the Lagrangian density for this continuum model being

$$\tilde{\mathscr{L}}(x) = \rho(x) ij \, \text{Tr}[\sigma_3 g(x)^{-1} \dot{g}(x)] \, ; \qquad x \equiv (\mathbf{x}, t) \qquad (C.3)$$

We now devise criteria for the construction of ρ:

A) We are interested in developing minimal Lagrangians which do not contain more fields than g. We therefore require that ρ depends on g alone.

B) We shall take the conservation of the total number of spins as our second requirement on ρ. This means that

$$N = \int d^2x \rho(x)$$

is a constant of the motion and motivates the assumption that ρ is the time component J^0 of a conserved current J^μ:

$$\partial_\mu J^\mu = 0 \, . \qquad (C.4)$$

ρ or J^0 is allowed to be negative. This is because ρ has at times an interpretation as the difference in the number density of solitons and antisolitons, which can be negative.

There is a natural class of currents J_μ which fulfill these requirements independently of dynamical details. They can be constructed as follows. Let

$$A_\mu = \frac{i}{2} \text{Tr}[\sigma_3 g^{-1} \partial_\mu g] \tag{C.5}$$

$$F_{\mu\nu} = \partial_\mu A_\nu - \partial_\nu A_\mu .$$

Then

$$J^\mu = K^\mu + \frac{1}{j} \frac{\theta}{4\pi} \epsilon^{\mu\nu\lambda} \frac{F_{\nu\lambda}}{4\pi} \tag{C.6}$$

where K^μ is any function of x fulfilling $\partial_\mu k^\mu = 0$. For instance if K^0 is time independent, we are free to take **K** to be zero.

In (C.6) we have assumed that $j \neq 0$. This is permissible since $\mathcal{L}$ is zero if j is zero. We have also chosen the factors in the second term so that the elementary soliton is a boson (fermion) if $\theta = 0 \mod 2\pi$ $(= \pi \mod 2\pi)$.

An interpretation of K^μ and its Lagrangian is obtained from the choice $K^0 =$ constant, **K** = 0. The Lagrangian density (with $\theta = 0$) is then (constant) $\times 2jA_0$. It describes noninteracting spins uniformly distributed on a plane. For a uniform spin lattice, this density function ρ is obtained simply by counting the number of spins per unit area. More general choices of K^0 may therefore be regarded as describing nonuniformly distributed spins. They are not translationally invariant systems and are not of interest to us.

The interpretation of the second term in (C.6) is somewhat different. Let us introduce a "spin" vector **S** of unit length ($\mathbf{S} \cdot \mathbf{S} = 1$) using the definition

$$g \sigma_3 g^{-1} = \sigma_\alpha S_\alpha . \tag{C.7}$$

Then

$$F_{12} = \frac{1}{2} \epsilon^{ijk} S_i \frac{\partial S_j}{\partial x^1} \frac{\partial S_k}{\partial x^2} = \frac{1}{2} \mathbf{S} \cdot \left[\frac{\partial \mathbf{S}}{\partial x^1} \times \frac{\partial \mathbf{S}}{\partial x^2} \right] . \tag{C.8}$$

But the standard volume form on the two-sphere $\mathbf{S}^2 = \{\mathbf{n} | \mathbf{n} \cdot \mathbf{n} = 1\}$ (with 4π as the total volume) is just $\epsilon_{ijk} n^i dn^j \wedge dn^k / 2$. Thus the charge N from the second term measures the degree of the map $\mathbf{x} \to \mathbf{S}(\mathbf{x})$ from the spatial slice to the two-sphere of spin space and the corresponding ρ is now the density of this topological charge. One imposes boundary conditions on g and hence **S** which permit us to regard the spatial slice as compact when considering this map [34]. N is then integer valued upto a constant

$$N = \frac{n\theta}{4\pi j}, \quad n \in \mathbb{Z} . \tag{C.9}$$

This density function is thus sensitive to the coherence of spins at distinct **x** and seems particularly ideal to describe solitonic and other collective excitations of spins. Since it is sensitive to the relative orientations of spins at distinct **x** and flips in sign if the target S^2 is traversed with different orientations when a spatial area is traversed, there is no reason for F_{12} to have a fixed sign. This is in contrast to the choice $\rho = K^0 =$ constant. If we now associate a positive ρ with a collective excitation and identify it with a quasiparticle density, then for $\rho < 0$ we must regard $|\rho|$ as the quasiantiparticle density.

W may remark here that F_{12} can be interpreted as the continuum limit of the P and T violating order parameter ($=\mathbf{S}_i \cdot (\mathbf{S}_j \times \mathbf{S}_k)$) of Wilczek et al. [179].

These considerations show that the two terms in J^μ describe rather different physical situations. It seems therefore reasonable to retain only one of the two terms in J^0 in a preliminary exploration. As our interest is in a model with correlations, we identify J^0 with the second term in (C.6). Such an identification is also motivated by the fact that it is the associated model which has interesting symmetry properties and topological features. With this identification, our continuum Lagrangian is

$$L = \int d^2x \mathscr{L}$$

$$\mathscr{L} = \frac{\theta}{4\pi} \frac{F_{12}}{\pi} A_0 \;.$$

(C.10)

It is worthwhile to mention here that this Lagrangian is equivalent to one of the terms in the expansion of the $\mathbb{C}\mathrm{P}^1$ Chern-Simons term $\mathscr{L}_{\mathrm{CS}}$. $\mathscr{L}_{\mathrm{CS}}$ is constructed from the connection field A_μ and reads

$$\mathscr{L}_{\mathrm{CS}} = \frac{\theta}{12\pi^2} \epsilon^{\mu\nu\lambda} A_\mu F_{\nu\lambda} \;. \tag{C.11}$$

(It is this term (with $\theta = \pi$) which was included in the $\mathbb{C}\mathrm{P}^1$ Lagrangian density by Polyakov to convert the $\mathbb{C}\mathrm{P}^1$ soliton to a spin $1/2$ particle [147]. See also Sec. 7.)

It was shown in Ref. [34] (and as we have briefly discussed above) that the model described by (C.10) has solitonic excitations. These solitons can have fractional intrinsic angular momentum and hence this model is in this respect similar to the abovementioned $\mathbb{C}\mathrm{P}^1$ model with the Chern-Simons term. It was further shown in Ref. [34] that this topological model resembles a system of anyons in the limit of several point solitons and hence has features characteristic of the mean field approach to the anyon gas. We will not discuss this model further here, and refer the reader to Ref. [34] for additional details.